Next Generation Sequencing in Forensic Science

Next Generation Sequencing in Forensic Science
A Primer

Kelly M. Elkins and Cynthia B. Zeller

CRC Press
Taylor & Francis Group
Boca Raton London New York

CRC Press is an imprint of the
Taylor & Francis Group, an **informa** business

First edition published 2022
by CRC Press
6000 Broken Sound Parkway NW, Suite 300, Boca Raton, FL 33487-2742

and by CRC Press
2 Park Square, Milton Park, Abingdon, Oxon, OX14 4RN

ISBN: 978-0-367-47893-3 (hbk)
ISBN: 978-1-032-07204-3 (pbk)
ISBN: 978-1-003-19646-4 (ebk)

DOI: 10.4324/9781003196464

Typeset in Minion
by codeMantra

To Emily, Julie, Madeleine, Kay and Sara and
all future scientists and teachers

Contents

Foreword   xi
Preface   xiii
Acknowledgments   xv
Authors   xvii
List of Figures   xix
List of Tables   xxiii
List of Credits   xxv
List of Abbreviations   xxvii

**1   History of DNA-Based Human Identification in
     Forensic Science**   1

   1.1   Introduction   1
   1.2   Application of DNA Sequencing to Human DNA   1
   1.3   History of DNA Typing   2
   1.4   Next Generation Sequencing for Forensic DNA Typing   8
   1.5   Conclusion   10
   Questions   10
   References   11

2   History of Sequencing for Human DNA Typing   13

   2.1   Introduction   13
   2.2   Common Chemistries Used in Sequencing Applications   13
         2.2.1   Chain Termination Sequencing   13
         2.2.2   Pyrosequencing   14
         2.2.3   Sequencing by Ligation   16
   2.3   Detection Techniques   17
         2.3.1   Fluorescence   17
         2.3.2   Pyrosequencing   19
         2.3.3   Ion Detection   19
   2.4   Sequencing Platforms   19
         2.4.1   First-Generation Sequencing Techniques   19
                 2.4.1.1   Sanger Sequencing   19
                 2.4.1.2   SNaPShot Sequencing   20
                 2.4.1.3   Pyrosequencing   21

	2.5	Massively Parallel Sequencing	23
		2.5.1 Reversible Chain Termination MPS Platforms	23
		2.5.2 Ion Detection Platforms	23
		2.5.3 Sequencing by Ligation Platforms	24
		2.5.4 Single Base Extension Platforms	25
		2.5.5 Third-Generation Platforms	25
	2.6	NGS Instruments Adopted for Forensic Science	25
		Questions	28
		References	28

**3   Sample Preparation, Standards, and Library
       Preparation for Next Generation Sequencing   31**

	3.1	Overview of the NGS Sample Preparation Process	31
	3.2	Sample Handling and Processing	31
	3.3	DNA Extraction	32
	3.4	DNA Quantitation	34
	3.5	Library Preparation	35
	3.6	Library Purification and Normalization	39
	3.7	Multiplexing and Denaturation	41
		Questions	42
		References	42

4   Performing Next Generation Sequencing   47

	4.1	Performing Next Generation Sequencing	47
	4.2	Verogen MiSeq FGx® Sequencing	47
	4.3	ThermoFisher Ion Torrent™ and Ion PGM Sequencing	53
	4.4	The Next Step	54
		Questions	54
		References	55

**5   Next Generation Sequencing Data Analysis
       and Interpretation   57**

	5.1	NGS Data Analysis	57
	5.2	Verogen Universal Analysis Software	58
	5.3	ThermoFisher Converge Software	69
	5.4	Phenotype Analysis Using the Erasmus Server	74
	5.5	Other Sequence Analysis Software	77
	5.6	Additional Tools for Mixture Interpretation	78
	5.7	Other NGS Sequence Data Analysis Tools	79
	5.8	NGS Validation and Applications	80
		Questions	83
		References	83

6   Next Generation Sequencing Troubleshooting   87

    6.1   Troubleshooting NGS Sequencing   87
    6.2   Troubleshooting MiSeq FGx Instrument Failure   87
    6.3   Troubleshooting MiSeq FGx Run Failure   89
    6.4   Troubleshooting Ion Series Run Failure   92
    Questions   94
    References   94

**7   Mitochondrial DNA Typing Using Next
Generation Sequencing   95**

    7.1   Introduction to Mitochondrial DNA Typing   95
    7.2   The Sequence of the Mitochondrial Chromosome   96
    7.3   Mitochondrial DNA Typing Methods   98
    7.4   Mitochondrial DNA Typing Using Next Generation
          Sequencing   98
    7.5   Mitochondrial Sequence Data Interpretation and
          Reporting   102
    7.6   Recent Reports of Mitotyping Using NGS for Forensic
          Applications   107
    7.7   Mitochondrial Sequence Data and Databases   108
    Questions   109
    References   109

**8   Microbial Applications of Next Generation
Sequencing for Forensic Investigations   117**

    8.1   Introduction to Microbial DNA Profiling   117
    8.2   Why NGS?   118
    8.3   The Human Microbiome Project   118
    8.4   Sampling and Processing   118
    8.5   NGS Methodology in Microbial Forensics   119
    8.6   Results from the Human Microbiome Project   120
    8.7   HMP Applications for Forensic Science   121
    8.8   NGS Applications in Geolocation, Autopsy, PMI, and
          Lifestyle Analysis   125
    8.9   Bioinformatic Approaches and Tools   126
    8.10 Bioforensics and Biosurveillance   127
    8.11 Infectious Disease Diagnostics   128
    8.12 NGS Applications in Archeology   129
    8.13 Summary of NGS Microbial Sequencing Applications
          in Forensic Investigation   129
    Questions   130
    References   130

**9   Body Fluid Analysis Using Next Generation
     Sequencing**   **137**

9.1   Introduction   137
9.2   Epigenetic-Based Tissue Source Attribution   137
9.3   mRNA-Based Tissue Source Attribution   139
9.4   MicroRNA Analysis   140
9.5   The Future of Body Fluid Assays   141
      Questions   141
      References   142

**10   Conclusions and Future Outlook of Next
      Generation Sequencing in Forensic Science**   **145**

10.1 NGS Is Here   145
10.2 Why NGS?   146
10.3 Ongoing Challenges of Adopting NGS for Forensic
      Investigations   147
10.4 Early Successes of NGS in Forensic Cases   152
10.5 Summary   154
      Questions   154
      References   154

Index   **159**

Foreword

Next Generation Sequencing (NGS) for the longest time was considered "the future of forensic DNA analysis." However, it is rapidly becoming a powerful tool in forensic DNA laboratories today for short tandem repeat (STR) typing, single nucleotide polymorphism (SNP) typing, and mitochondrial DNA. NGS technology can be a game changer for helping to solve crimes, create investigational leads, and solve complex ancestry cases. While capillary electrophoresis (CE) remains routine in forensic DNA analysis, the introduction of NGS to the forensic DNA field allows an alternative solution for the analysis of challenging forensic casework samples. In recent years, commercial companies have been releasing ready-to-use chemistries and protocols that can easily be incorporated into existing workflows in forensic DNA laboratories.

A clear advantage of using NGS for DNA typing is the sheer amount of additional information that can be obtained from the same sample input that is currently used with CE technologies. This includes additional loci available across autosomal, Y-, and X-STR markers as well as multiplexing multiple marker types within a single amplification. For example, the largest commercially available CE kit (at the present time) is the Investigator Argus Y-28 kit from Qiagen that includes 28 Y-STR markers in one system. The Forensic DNA Signature Prep kit from Verogen combines 27 autosomals, 7 X-STRs, and 24 Y-STRs in addition to 94 identity, 56 ancestry, and 22 phenotypic-informative SNPs for over 200 markers in a single multiplex. The additional information also includes sequence variations within markers that could potentially aid the resolution of complex cases with degraded or low amounts of DNA, assist with mixture deconvolution, help resolve kinship scenarios, and strengthen the statistics in population databases.

This textbook takes you through the history of forensic DNA-based human identification to include a variety of techniques such as VNTR, RFLP, STR, and SNP DNA typing and progresses to the history of sequencing in the forensic DNA community. Readers will dive into the entire process of NGS, to include sample and library preparation with a variety of commercial chemistries, setting up and performing sequencing with two different instruments, data analysis, interpretation, and troubleshooting issues that can occur. In addition, it covers a multitude of marker sets to include SNPs and mitochondrial DNA, and several NGS applications such as microbial

DNA and body fluid analysis. Readers will learn about future considerations and applications of this rapidly emerging technology.

The coauthors, Dr. Kelly Elkins and Dr. Cynthia Zeller, are both exceptional professors and researchers at Towson University. I've had the pleasure of knowing both for many years and can say with certainty that they have vast knowledge, experience, and enthusiasm for NGS and all it has to offer the forensic DNA community. They review the NGS process in detail and have written one of the first books to prepare practitioners to utilize and implement this important technology into their laboratories for forensic casework. In addition, this resource will aid in the education of future forensic scientists as they want to learn more about this ever-evolving technology.

Carolyn (Becky) Steffen, M.S., Research Biologist at the National Institute of Standards and Technology
Material Measurement Laboratory, Biomolecular Measurement Division,
Applied Genetics Group
June 2021

Preface

We have both taught Forensic Biology for several years but introduced our first course on next generation sequencing to our students only two years ago. This endeavor was supported by our Provost, Dean, Department Chair, and Program Director who strongly advocated for and financially supported instrument upgrades, reagents and consumables, and time for us to develop two new courses. Our department and program have added courses centered on course-based research using mitochondrial and autosomal DNA sequencing for undergraduate and graduate students. We have used the primary literature, books, videos, recorded lectures, and conference lectures and posters as our sources as we developed the courses. Our goal in this project was to bring the essential content and references together in one place for individuals interested in learning about next generation sequencing and its forensic applications. We hope that professional forensic DNA analysts, laboratory directors, and educators at all levels and their students will find this book to be an introduction to next generation sequencing in forensic science. We look forward to hearing from our readers.

Kelly M. Elkins
Cynthia B. Zeller
February 2021

Acknowledgments

We are fortunate to have received support from many people to help us succeed in this endeavor. From Towson University, Mark Profili, Program Director for our Forensic Programs at Towson University (TU), spearheaded the effort to introduce NGS in our forensic programs, and we thank him for his vision and leadership. We thank Ryan Casey, Chair of the Chemistry Department, for supporting us in adding this capability to our forensic program and allowing us flexibility in scheduling and workload. Vonnie Shields, Interim Dean, and David Vanko, Dean, supported us by funding two proposals for the new courses through the Jess and Mildred Fisher College of Science and Mathematics Fisher Endowment Funds. The TU Provost's Budget Office funded the mitochondrial DNA analysis software upgrade. Brian Masters was the Principle Investigator for the awarded NSF grant that enabled TU to acquire a MiSeq in 2013, and he willingly agreed to the FGx upgrade and additional users we represent – we couldn't have done this without him. Thanks to Dana Kollmann for the wonderful collaboration and opportunity to put the science to work. We were welcomed by Laura Gough and Matt Hemm into their Towson University Research Enhancement Program (TU-REP) generously funded by the Howard Hughes Medical Foundation (HHMI). They provided both financial support for course supplies and also excellent professional development training led by Rommel Miranda to us as we engaged in offering course-based undergraduate research experience (CURE) courses. We thank Michelle Snyder, Chris Ouferio, Barry Margulies, Larry Wimmers, Vanessa Beauchamp, and Elana Ehrlich for excellent discussions in the TU-REP community. Thanks to Adam Klavens for working with us to conduct many of the experiments and troubleshooting that guide our explanations in this book and for providing helpful comments and suggestions on the manuscript. Hirak Ranjan Dash reviewed several chapters and we appreciate his feedback. We are also grateful to the teams at Verogen – including Danny Hall, Kristi Kim, and Melissa Kotkin – and Qiagen – including Mary Jones-Dukes and Mark Guillano – for teaching us and guiding us in troubleshooting and Becky Steffen of NIST for invaluable discussions. Finally, this book would not have been possible without our TU students who are the reason for our new courses, which laid the groundwork for this book.

Thanks to Mark Listewnik, acquisitions editor, who immediately supported this project and rapidly sent our proposal for review and found excellent reviewers who provided invaluable suggestions. It is always a plea-sure to work with you. Thank you also to Katie Horsfall at CRC Press for administrative support.

We look forward to hearing from our readers and forensic analysts who introduce this process into their daily workflow.

I (Kelly) never thought I'd write a book – and this marks my third, not counting several additional book chapters. My husband and children con-tinue to support me, cheer me on, and listen to me talk through my projects. I appreciate them all more than words can say.

Authors

Kelly M. Elkins, PhD, is an Associate Professor of Chemistry at Towson University and a founding editor-in-chief of the *Journal of Forensic Science Education.* She has authored the books *Forensic DNA Biology: A Laboratory Manual* and *Introduction to Forensic Chemistry,* in addition to ten invited book chapters and more than thirty-five journal papers on her research in journals, including the *Journal of Forensic Sciences, Analytical Biochemistry, Drug Testing and Analysis,* and *Medicine, Science and the Law.* She has taught courses in forensic biology and forensic chemistry under various course numbers at four colleges and universities since 2006. She is an active member of the American Chemical Society and a Fellow of the American Academy of Forensic Sciences. She is a member of the Council of Forensic Science Educators and served as its President in 2012. She is a member of the ACS Ethics Committee and co-wrote the 2018 ACS Exams Institute Diagnostic of Undergraduate Chemical Knowledge (DUCK) exam. Her research has been funded by the Forensic Sciences Foundation, NSF, NIH, Maryland TEDCO, and ACS. She enjoys communicating science in the classroom, via outreach activities, in interviews, and on television. She is the editor for three books in production.

Cynthia B. Zeller, PhD, is an Associate Professor of Chemistry at Towson University. She has taught several forensic biology and DNA typing courses at Frederick Community College and Towson University for over fifteen years. After completing postdoctoral appointments in the School of Medicine at Johns Hopkins University for six years, she served as a Serologist and DNA Analyst at the Maryland State Police Forensic Science Division for six years. She is a member of the Mid-Atlantic Association of Forensic Scientists. She has published ten scientific publications and has delivered more than hundred conference and seminar presentations. Her work has been published in the *Journal of Forensic Sciences, Fibrogenesis Tissue Repair,* and *The American Journal of Physiology,* and it has been funded by the National Institutes of Justice. This is her first book.

List of Figures

Figure 1.1   Comparison of a single nucleotide polymorphism and short tandem repeat. (a) rs12913832 is a SNP near the OCA2 gene that has been shown to be linked to blue (G, G) or brown (A, A or A, G) eye color and is used in phenotype prediction assays. (b) DYS19 is an STR (NCBI Accession number X77751) with a variable number of repeats   7

Figure 1.2   Data collected using (a) a modern six-dye kit using CE and (b) NGS. (Courtesy of Adam Klavens.)   9

Figure 1.3   Decreasing cost per megabase of DNA sequence over time. (https://search.creativecommons.org/photos/ a2a51821-36b0-4432-9a8b-0e86317fca5c.)   10

Figure 2.1   DNA synthesis with growing DNA template and pyrophosphate byproduct. (Michał Sobkowski, CC BY 3.0 license.)   14

Figure 2.2   Comparison of Sanger sequencing and NGS processes. (Dale Muzzey, Eric A. Evans, Caroline Lieber, CC BY 4.0 license. https://www.ncbi.nlm.nih.gov/pmc/articles/ PMC4633438/figure/Fig1/.)   15

Figure 2.3   Overview of several DNA sequencing techniques with the principle of (a) Sanger sequencing, (b) pyrosequencing (e.g., 454), (c) em-PCR (e.g., 454, SOLiD® and Ion Torrent™) and (d) bridge amplification/cluster PCR (e.g., Solexa). (Brigitte Bruijns, Roald Tiggelaar, and Han Gardeniers, CC BY NC ND 4.0 license. https://www.ncbi. nlm.nih.gov/pmc/articles/PMC6282972/figure/elps6707- fig-0002/?report=objectonly.)   17

Figure 2.4   Overview of several DNA sequencing techniques with the principle of (a) sequencing by ligation (SBL, e.g., SOLiD®), (b) ion detection (e.g., Ion Torrent™), (c) zero-mode waveguides (ZMWs, e.g., PacBio®), and (d) nanopores (e.g., Oxford Nanopore). (Brigitte Bruijns,

Roald Tiggelaar, and Han Gardeniers, CC BY NC ND
4.0 license. https://www.ncbi.nlm.nih.gov/pmc/articles/
MC6282972/figure/elps6707-fig-0003/?report=objectonly.)   18

Figure 2.5   A sample Sanger sequence read.   21

Figure 2.6   Qiagen Q48 pyrosequencing instrument.   22

Figure 2.7   Verogen MiSeq FGx instrument.   24

Figure 3.1   Overview of steps in the NGS sample preparation process.   32

Figure 3.2   Comparison of steps in the library preparation
workflows for three products.   37

Figure 3.3   Agarose gel of ForenSeq library preparation amplicons
(From left to right: Lane 1: Trackit 50 bp ladder with
bright bands at 350/800/2500 bp, Lanes 2–6: DNA
standards, Lane 7: NTC, Lane 8: DNA standard from
10-month-old library prep). (Courtesy of Adam Klavens.)   40

Figure 3.4   QIAcel graph of PCR amplicon for pyrosequencing.   41

Figure 4.1   Setting up a sequencing run on the MiSeq FGx (RUO or
Forensic Use).   48

Figure 4.2   Wash screen on MiSeq FGx.   48

Figure 4.3   Preparing a new run in the Verogen Universal Analysis
Software.   49

Figure 4.4   Micro (left) and standard (right) MiSeq flow cells.   50

Figure 4.5   Sequencing in process on the MiSeq FGx.   51

Figure 4.6   MiSeq FGx sequencing run completion viewed in UAS.   52

Figure 5.1   MiSeq FGx run metrics for a successful sequencing run.   58

Figure 5.2   Passing HSC in MiSeq FGx sequencing run.   63

Figure 5.3   ForenSeq sequencing run negative control with no alleles.   63

Figure 5.4   UAS reads versus length graph for 2800M.   64

Figure 5.5   Full profile for 2800M using ForenSeq library prep.   64

Figure 5.6   Sample comparison in UAS for 9948 at two input
concentrations.   65

Figure 5.7   D7S820 locus displaying stutter and peak imbalance
for a sample.   65

Figure 5.8   Relatively Balanced rs12913832 SNP reads for a sample.   66

Figure 5.9   Imbalanced rs1413212 SNP reads per sample.   67

Figure 5.10   UAS phenotype estimate for 2800M.   68

Figure 5.11   GlobalFiler NGS STR panel v2 data viewed with
                   Converge software.   72

Figure 5.12   Eye color prediction tree using SNPs.   74

Figure 5.13   Erasmus SNP input for K562 prediction of phenotype.   75

Figure 5.14   Erasmus K562 phenotype prediction.   76

Figure 5.15   UAS biogeographical ancestry and phenotype
                   prediction for K562 prepared with ForenSeq.   76

Figure 5.16   Number of alleles for Y-STRs analyzed using
                   CE and NGS.   81

Figure 5.17   Number of alleles for Y STRs analyzed using
                   CE and NGS.   82

Figure 6.1   MiSeq FGx Y-stage home error.   88

Figure 6.2   MiSeq FGx camera focus error.   89

Figure 6.3   MiSeq FGx run failure viewed in UAS.   91

Figure 7.1   Variation between SNP 73 in HL-60 as compared
                   to the rCRS.   103

Figure 7.2   Reads per sample.   103

Figure 7.3   Insertion at 315.1 in the HL-60 standard as compared to
                   the rCRS.   104

Figure 7.4   Total reads and calls for a sample at three
                   concentrations (1, 5, and 100 pg) at position 489.   105

Figure 7.5   A purple warning flag indicates that the number of
                   reads for position 523 is below the IT in a 1 pg sample.   105

Figure 8.1   Bias in sequencing in the gut microbiome using Sanger,
                   454, SOLiD and Shotgun-SOLiD sequencing methods.
                   (Suparna Mitra, Karin Forster-Fromme, Antje Damms-
                   Machado, Tim Scheurenbrand, Saskia Biskup, Daniel H.
                   Huson, Stephan C. Bischoff (CC 2.0). https://pubmed.
                   ncbi.nlm.nih.gov/24564472/.)   122

Figure 8.2   Normalized comparison between 16S samples obtained
                   using three technologies: "Sanger," "16S-454," and
                   "16S-SOLiD" datasets. Normalized comparison result

obtained using MEGAN for "Sanger"-dataset (blue), "16S-454" dataset (cyan), and "16S-SOLiD" dataset (magenta) without considering "No hits" node. The tree is collapsed at "family" level of NCBI taxonomy. Circles are scaled logarithmically to indicate the number of summarized reads. (Suparna Mitra, Karin Förster-Fromme, Antje Damms-Machado, Tim Scheurenbrand, Saskia Biskup, Daniel H. Huson, Stephan C. Bischoff (CC 2.0). https://pubmed.ncbi.nlm. nih.gov/24564472/#&gid=article-figures&pid=figure-3- uid-2.)   123

Figure 8.3   Summary of forensic applications of microbial NGS.   130

Figure 9.1   Pyrograms resulting from a vaginal epithelial sample analyzed with the Body Fluid Identification Multiplex. Vaginal epithelia is characterized by moderate methylation in the BCAS4 assay (a), hypomethylation in the cg06379435 assay (b), moderate methylation in the VE_8 assay (c), and hypermethylation in the ZC3H12D assay (d). The combination of multiple body fluid assays in a single reaction allows for higher accuracy in body fluid identification while reducing sample consumption and costs. (Courtesy of Quentin Gauthier.)   140

Figure 10.1   Summary of challenges of adopting NGS for forensic investigations.   152

List of Tables

Table 1.1 Brief History of Some Notable Advances in Forensic Biology 2

Table 1.2 Core Str Loci for Several Countries and Unions 6

Table 2.1 NGS Read Length, Run Time, and Per Base Cost 26

Table 2.2 Advantages and Disadvantages of Various Sequencing Approaches 27

Table 3.1 Manufacturer's Recommended DNA Input Quantity for NGS 35

Table 3.2 Comparison of Steps and Time Required for Library Preparation for Kits from Commercial Suppliers 36

Table 3.3 i7 Index Labels and Sequences 38

Table 3.4 i5 Index Labels and Sequences 38

Table 5.1 ForenSeq and Precision ID Target Autosomal and Sex Chromosomal STR Loci 60

Table 5.2 Key Notes for Inputting UAS Calls to Erasmus Server 76

Table 6.1 Troubleshooting the MiSeq FGx Instrument and Sequencing Runs 93

Table 7.1 Variations between the CRS and rCRS Mitochondrial Chromosome Sequences 96

Table 7.2 Frequently Probed Mitochondrial DNA SNP Positions in the Variable Regions (HVI, HVII, and HVIII) and Outside the Control Region (Other) 97

Table 9.1 Genetic Markers Identified Body Fluid Identification Using Pyrosequencing 139

List of Credits

Figure	Credit Line	Caption
Figure 1.2	Adam Klavens	Data collected using (a) a modern 6-dye kit using CE and (b) NGS
Figure 1.3	@dullhunk (CC BY 2.0)	Decreasing cost per megabase of DNA sequence over time
Figure 2.1	Michał Sobkowski, (CC BY 3.0)	DNA synthesis with growing DNA template and pyrophosphate byproduct
Figure 2.2	Dale Muzzey, Eric A. Evans, Caroline Lieber (CC BY 4.0)	Comparison of Sanger sequencing and NGS processes
Figure 2.3	Brigitte Bruijns, Roald Tiggelaar, and Han Gardeniers (CC BY NC ND 4.0)	Overview of several DNA sequencing techniques with the principle of (A) Sanger sequencing, (B) pyrosequencing (e.g., 454), (C) em-PCR (e.g., 454, SOLiD® and Ion Torrent™), and (D) bridge amplification/cluster PCR (e.g., Solexa).
Figure 2.4	Brigitte Bruijns, Roald Tiggelaar, and Han Gardeniers, CC BY NC ND 4.0	Overview of several DNA sequencing techniques with the principle of (A) sequencing by ligation (SBL, e.g., SOLiD®), (B) ion detection (e.g., Ion Torrent™), (C) zero-mode waveguides (ZMWs, e.g., PacBio®), and (D) nanopores (e.g., Oxford Nanopore).
Figure 2.6	Madeleine Phillips	Qiagen Q48 pyrosequencing instrument
Figure 2.7	Madeleine Phillips	Verogen MiSeq FGx instrument
Figure 3.3	Adam Klavens	Agarose gel of ForenSeq library preparation amplicons (From left to right: Lane 1: Trackit 50 bp ladder with bright bands at 350/800/2500 bp, Lanes 2–6: DNA standards, Lane 7: NTC, Lane 8: DNA standard from 10 month old library prep)
Figure 4.4	Madeleine Phillips	Micro and standard MiSeq flow cells
Figure 5.11	Hirak Ranjan Dash	GlobalFiler NGS STR panel v2 data viewed with Converge software
Figure 5.13	Adam Klavens	Erasmus SNP input for K562 prediction of phenotype
Figure 5.14	Adam Klavens	Erasmus K562 phenotype prediction
Figure 8.1	Suparna Mitra, Karin Förster-Fromme, Antje Damms-Machado, Tim Scheurenbrand, Saskia Biskup, Daniel H. Huson, Stephan C. Bischoff (CC 2.0)	Bias in sequencing in the gut microbiome using Sanger, 454, SOLiD, and Shotgun-SOLiD sequencing methods

Figure	Credit Line	Caption
Figure 8.2	Suparna Mitra, Karin Förster-Fromme, Antje Damms-Machado, Tim Scheurenbrand, Saskia Biskup, Daniel H. Huson, Stephan C. Bischoff (CC 2.0)	Normalized comparison between 16S samples obtained using three technologies: "Sanger," "16S-454," and "16S-SOLiD" datasets. Normalized comparison result obtained using MEGAN for "Sanger"-dataset (blue), "16S-454" dataset (cyan) and "16S-SOLiD" dataset (magenta) without considering "No hits" node. The tree is collapsed at "family" level of NCBI taxonomy. Circles are scaled logarithmically to indicate the number of summarized reads.

List of Abbreviations

A	Adenine
AAFS	American Academy of Forensic Sciences
AFDIL	Armed Forces DNA Identification Laboratory
aiSNP	Ancestry informative single nucleotide polymorphism
ALFRED	ALlele frequency database
AMEL	Amelogenin
AMP	Adenosine monophosphate
AN	Allele number
APS	Adenosine-5′-phosphosulfate
AQME	AFDIL-QIAGEN mtDNA Expert
aSTR	Autosomal short tandem repeat
AT	Analytical threshold
BGA	Biogeographical ancestry
bp	Base pairs
C	Cytosine
cDNA	Complementary DNA
CE	Capillary electrophoresis
CODIS	Combined DNA Index System
CNV	Copy number variation
CpG	Cytosine phosphate guanine
CRS	Cambridge Reference Sequence
CSF1PO	c-fms proto-oncogene for CSF-1 receptor gene
ddNTP	2′, 3′-dideoxyribonucleotide triphosphate
DNA	Deoxyribonucleic acid
dNTP	2′-deoxyribonucleotide triphosphate
dsDNA	Double-stranded DNA
EDTA	Ethylene diamine tetraacetic acid
ELISA	Enzyme-linked immunosorbent assay
ENFSI	European Network of Forensic Science Institutes
ESS	European Standard Set
FBI	Federal Bureau of Investigation
FEPAC	Forensic Science Education Programs Accreditation Commission
FGA	Alpha fibrinogen gene
FTA	Flinders Technology Associates

G	Guanine
GATK	Genome analysis toolkit
GI	Gastrointestinal
GITAD	Ibero American Scientific Working Group on DNA Analysis
HID	Human identification
HIPPA	Health Insurance Portability and Accountability Act of 1996
HipSTR	Haplotype inference and phasing for short tandem repeats
HMP	Human microbiome project
HRM	High resolution melt
HSC	Human sequencing control
HV	Highly/hyper variable region
ID	Identity
IGV	Integrative genomics viewer
iiSNP	Identity informative single nucleotide polymorphism
INDEL	Insertion/deletion
ISFET	Ion-sensitive field effect transistor
ISP	Ion sphere particles
IT	Interpretation threshold
LDA	Linear discriminant analysis
LR	Likelihood ratio
miRNA	MicroRNA
MPS	Massively parallel sequencing
mRNA	Messenger RNA
mtDNA	Mitochrondrial DNA
MUSCLE	MUltiple sequence comparison by log-expectation
MyFLq	My-forensic-loci-queries
NCBI	National Center for Biotechnology Information
NDIS	National DNA Index System
NGC	Next generation sequencing confirmation
NGS	Next generation sequencing
NIH	National Institutes of Health
NIST	National Institute of Standards and Technology
NTC	No template control
OF	Off-ladder alleles
ORF	Open reading frame
OSAC	Organization of Scientific Area Committee
OTU	Operational taxonomic unit
PCA	Principal component analysis
PCI or PCIA	Phenol-chloroform-isoamyl alcohol
PCR	Polymerase chain reaction
PE	Probability of exclusion

PGM	Personalized genomic machine
PHR	Peak height ratio
PIRANHA	Programme d'interprétation résultats d'analyses NGS hautement amélioré
PLS	Partial least squares
PMI	Post-mortem interval
PNL	Pooled, normalized libraries
POP	Performance-optimized polymer
PPE	Personal protective equipment
PPI	Pyrophosphate
pSNP	Phenotype single nucleotide polymorphism
Q	Quality score
QA	Quality assurance
QC	Quality control
OTUs	Operational taxonomic units
rCRS	Revised Cambridge reference sequence
RFID	Radiofrequency identification
RFLP	Restriction Fragment Length Polymorphism
RFU	Relative fluorescence units
RMNE	Random man not excluded
RMP	Random match probability
RNA	Ribonucleic acid
rRNA	Ribosomal RNA
RSB	Resuspension buffer
RT-PCR	Reverse transcriptase-polymerase chain reaction
RUO	Research use only
SAM	Sequence alignment map
SAP	Shrimp alkaline phosphatase
SARS-CoV-2	Severe acute respiratory syndrome coronavirus 2
SBS	Sequencing by synthesis
SIDS	Sudden infant death syndrome
SMRT	Single molecule, real-time
SNP	Single nucleotide polymorphism
snRNA	Small nuclear RNA
SOLiD	Sequencing by oligonucleotide ligation and detection
SOP	Standard operating procedure
SRM	Standard reference material
ssDNA	Single-stranded DNA
SSR	Simple sequence repeat
STR	Short tandem repeat
SWG	Scientific Working Group
SWGDAM	Scientific Working Group on DNA Analysis Methods

T	Thymine
tDMR	Tissue-specific differentially methylated regions
TH01	Tyrosine hydroxylase 1 gene
TPOX	Thyroid peroxidase gene
TWG	Technical Working Group
UAS	Universal Analysis System
VCF	Variable call format
VNTR	Variable number tandem repeat
VPN	Virtual private network
VWA	von Willebrand factor gene
WGA	Whole genome amplification
WGS	Whole genome shotgun
WMS	Whole metagenome shotgun

History of DNA-Based Human Identification in Forensic Science

1

1.1 Introduction

Forensic biology is the application of serology and DNA typing methods for human, wildlife, pet, and plant identification using bone, teeth, hair, body fluids, and plant materials to help solve a crime. All of these materials, indeed all cellular material with the exception of red blood cells, contain deoxyribonucleic acid (DNA), the chemical whose double-stranded helical structure was elucidated in 1953 (Watson and Crick 1953, Franklin and Gosling 1953, Wilkins et al. 1953). Advances in forensic biology have resulted in tremendous capabilities for human identification and have reduced the reliance on and need for eyewitness accounts of crimes. A brief history of some of the notable advances in forensic biology is listed in Table 1.1.

1.2 Application of DNA Sequencing to Human DNA

The human genome sequence was reported in 2001 (International Human Genome Consortium 2001, Venter et al. 2001) and is comprised of DNA housed in the nucleus. This huge advance built on principles and technology that led to the sequencing of the bacteriophage phi X174 in 1977 (Sanger et al. 1977) and mitochondrial organelle chromosome in 1981 (Anderson et al. 1981). Whereas the human mitochondrial chromosome is circular in structure and is comprised of 16,569 base pairs (bp), the nuclear, or autosomal, genome consists of 3.2 billion bp packaged in twenty-two pairs of linear chromosomes supercoiled on histone proteins (Anderson et al. 1981, International Human Genome Consortium 2001, Venter et al. 2001). An additional set of chromosomes, X and Y, are sex chromosomes that make the total number of chromosomes in humans forty-six. Females have two X chromosomes, while males are characterized by having an X and a Y chromosome. Most of this book is focused on autosomal DNA typing and chapter 7 is focused on mitochondrial DNA typing. Over several years, the sequences of bacteria that have been used as foodborne pathogens and bioterror agents and others found in and on the human body have also been sequenced and used to solve questions in forensic cases; this is the focus of Chapter 8.

DOI: 10.4324/9781003196464-1   1

Table 1.1   Brief History of Some Notable Advances in Forensic Biology

Year	Advance
1953	Rosalind Franklin records X-ray autoradiographs of crystallized DNA fibers and deduces basic features including that the structure was helical with the phosphates on the outside and its basic dimensions of DNA strands and publishes with Raymond Gosling in *Nature*
1953	James Watson and Francis Crick solve three-dimensional structure of DNA from Franklin's X-ray crystallography data and publish it in *Nature*
1977	Frederick Sanger invents a method for DNA sequencing
1981	Anderson and team sequence human mitochondrial chromosome
1983	Kary Mullis invents polymerase chain reaction method as reported in *Science*
1985	Alec Jeffreys develops the first multi-locus DNA typing method and publishes it in *Nature*
1986	Human leukocyte antigen HLA-DQα multi-allelic locus was published for forensic DNA typing and used in the first US criminal case
1986	First application of Jeffrey's method to rape and murder cases in Leicestershire, England
1990	FBI establishes the Combined DNA Index System (CODIS) to allow for national DNA comparisons
1994	Promega introduces first commercial three loci STR typing kit targeting CSF1PO, TPOX, and TH01, named "CTT" using the first letter of each locus
1996	In *Tennessee v. Ware*, mitochondrial DNA typing was admitted for the first time in a US court.
1997	Applied Biosystems introduces the three-dye AmpFlSTR® Profiler Plus® PCR Amplification Kit for typing nine STRs and Amelogenin
1998	Applied Biosystems introduces the three-dye AmpFlSTR® COfiler® PCR Amplification Kit for typing six STRs and Amelogenin
2000	Promega introduces PowerPlex™ 16 System, the first commercial STR typing kit that targeted all thirteen CODIS loci as well as the ENFSI loci, Interpol loci, and GITAD loci in one PCR reaction
2001	Applied Biosystems introduces five-dye AmpFLSTR™ Identifiler™ PCR Amplification Kit targeting sixteen STR loci
2014	ThermoFisher introduces GlobalFiler™, the first twenty-four-plex, six-dye STR kit
2014	Promega introduces the PowerPlex® Fusion 24-locus STR DNA typing system
2016	Promega introduces PowerSeq™ 46GY kit
2017	Qiagen introduces the Investigator 24plex GO! and DS kits that target the CODIS and ESS loci
2018	Applied Biosystems introduces Precision ID NGS library prep kit
2017	Verogen introduces NGS ForenSeq™ library prep kit for forensic applications
2019	Precision ID and ForenSeq™ data approved for inclusion in CODIS
2019	First application of NGS in a Dutch criminal case.

1.3 History of DNA Typing

Analysis of the DNA sequence and repeat polymorphisms is referred to as DNA typing, or DNA profiling. DNA typing is used to determine the origin of forensic evidence and is applied to criminal, paternity, and missing persons cases. When forensic DNA typing was pioneered and adopted in the 1980s, relatively small segments of the genome were probed and used to differentiate the origin of samples (Gill et al. 1985). Even without sequencing the entire human genome, these DNA typing methods were shown to provide high statistical confidence that a stain or sample can be assigned as originating from a specific individual (Butler 2005). However, as sequencing capabilities and tools have improved, they are now being applied to forensic cases. DNA sequencing can overcome limitations of the now standard DNA typing methods. For example, monozygotic twins and body fluids can now be differentiated genetically. In addition, new, so-called next generation sequencing (NGS) methods can be used to determine the biogeographical ancestry and phenotype characteristics including eye color, skin tone, and hair color from fragments of human remains and trace body fluid or fingerprint sources.

It has been estimated that 30% of the human genome is comprised of repeated segments. The origins of forensic DNA typing began with restriction fragment length polymorphisms (RFLPs). The RFLP targets were sites with a variable number of tandem repeats (VNTRs). These sites are also known as minisatellite sequences and are comprised of long repeats of ten to one hundred nucleotide bases consisting of thousands of bases in total (Butler 2005). The first RFLP DNA test was developed by Sir Alec Jeffreys in 1984 and published in the journal *Nature* in 1985 (Gill et al. 1985). The test involved analysis of patterns from multiple RFLP loci. In RFLP, restriction enzymes were used to cut the DNA repeat region, and gel electrophoresis was used for separation and sizing. Following electrophoresis, the DNA was chemically denatured to separate the strands. The fragments were transferred to a nylon membrane and analyzed using a Southern Blot. The nylon membrane was serially treated with individual radioactive probes containing an oligonucleotide complementary to the target RFLP sequence. After the sequence hybridized with the target, the excess probes were washed off, the membrane was developed on an X-ray film, and bands appeared where the radioactive probe hybridized. The length of the VNTR was determined using a ladder consisting of DNA fragments of known lengths. DNA typing was a slow process. Analysis of a sample could take six to eight weeks. The use of radioactive probes for detection exposed analysts to radioactivity that added up over time. RFLP required relatively large quantities of high molecular weight and intact double-stranded DNA, so it was useless with trace, damaged, or degraded DNA samples. Finally, the gel bands were categorized into bins, or size groupings, rather than discrete base pair fragment sizes.

RFLP analysis was first employed in casework in 1986 in an investigation of the rape and murder of two girls in England (Butler 2005). Interestingly, the initial test results exonerated an innocent suspect. Additional testing led to the identification of the perpetrator, Colin Pitchfork. A few years later, a seven-probe RFLP assay was used to analyze DNA extracted from body fluids on intern Monica Lewinsky's blue dress in a 1993 sexual case involving then-U.S. President Bill Clinton (Butler 2005).

After the polymerase chain reaction (PCR) was invented by Kary Mullis in 1983 and published in *Science* in 1985 (Saiki et al. 1985), it became possible to amplify DNA targets prior to RFLP typing, although the length of the target RFLP sequences was not very conducive to PCR. Henry Erlich and colleagues developed the first PCR-based forensic, HLA DQα (DQA1), test in 1986 (Erlich et al. 1986), and it was used in a civil court case that year. In the case *People v. Pestinikas*, forensic scientist Edward Blake used a PCR-based DNA test to confirm that different autopsy samples originated from the same person. This was the first use of PCR-based DNA testing in the United States.

Short tandem repeats (STRs) are repeated elements in the nuclear genome. STRs, or simple sequence repeats (SSRs), are VNTR microsatellites with short repeats of only two to six base pairs. Even if the sequence is repeated tens of times, the overall length is much more amenable to PCR. Additionally, STRs are common in the genome, occurring approximately once per 10,000 bases (Butler 2005). STRs were first used for human identification for law enforcement and forensic purposes in 1992 after Thomas Caskey, professor at Baylor University in Texas, and colleagues published the first paper suggesting STRs for forensic DNA analysis in 1991 (Edwards et al. 1991). Although small gels were initially used to analyze STRs, large sequencing gels were eventually employed to enable better differentiation, and most STR separation today is performed using capillary electrophoresis (CE). Large polyacrylamide slab gels and capillary electrophoresis enable discrete sizing to a single base pair ending an important limitation of RFLP analysis.

Beginning in the 1990s, STR PCR primers were multiplexed so that they could be used to probe multiple STR loci simultaneously. Promega Corporation and Perkin-Elmer Corporation, in collaboration with Roche Molecular Systems, independently developed commercial kits for forensic DNA STR typing. The first commercially available multi-locus STR kit was the triplex "CTT" kit from Promega introduced in 1994 that enabled the determination of the number of repeats at the CSF1PO, TPOX, and TH01 loci using silver staining. Additional kits followed, each increasing the number of sites and/or varying the loci for identification. Applied Biosystems introduced the three-dye AmpFlSTR® Profiler Plus® PCR Amplification Kit for typing nine STRs and the amelogenin marker for sex determination in 1997 and the COfiler® PCR Amplification Kit in 1998. In these kits, the PCR primers were labeled with a fluorescent dye that enables amplicon detection and sizing with CE or a large slab gel. While STRs have been primarily used for DNA typing, biallelic loci, such as those located on the X and Y sex chromosomes, can also be employed for DNA typing and are used for sex typing via the amelogenin gene (Elkins 2013).

To enable interagency use of DNA typing data to solve criminal cases, the United States' Combined DNA Index System (CODIS) database was piloted by the FBI Laboratory in 1990, and the National DNA Index System (NDIS) was established by the DNA Identification Act of 1994 for law enforcement purposes. CODIS also includes a mitochondrial DNA (mtDNA) database. By 1998, all of the states in the United States enacted statutes requiring mandatory DNA testing for convicted felons. Promega introduced the PowerPlex™ 16 System in 2000. It was the first commercial STR typing kit that targeted all thirteen CODIS loci as well as the European Network of Forensic Science Institutes (ENFSI) loci, Interpol loci, and the Ibero American Scientific Working Group on DNA Analysis (GITAD) loci in one PCR reaction.

Beginning January 1, 2017, the number of loci required to upload an STR profile to CODIS increased from thirteen to twenty loci (Hares 2015). The current CODIS loci are CSF1PO, D3S1358, D5S818, D7S820, D8S1179, D13S317, D16S539, D18S51, D21S11, FGA, TH01, TPOX, vWA, D1S1656, D2S441, D2S1338, D10S1248, D12S391, D19S433, and D22S1045 and are listed in Table 1.2. Other nations and unions independently adopted their own sets of STR loci for forensic use. This is also shown in Table 1.2. The ThermoFisher GlobalFiler™ kit was introduced in 2014. The GlobalFiler™ kit includes PCR primers labeled with one of six dyes to amplify twenty-four loci simultaneously and covers all of the loci now required by the United States, the United Kingdom, and in the European Standard Set (ESS) (Table 1.1). The GlobalFiler™ Kit loci are D13S317, D7S820, D5S818, CSF1PO, D1S1656, D12S391, D2S441, D10S1248, D18S51, FGA, D21S11, D8S1179, vWA, D16S539, TH01, D3S1358, D2S1338, D19S433, DYS391, TPOX, D22S1045, SE33, ame-logenin, and a Y-specific insertion/deletion locus (Yindel). Similarly, the Promega PowerPlex® Fusion System 24-locus multiplex kit introduced in 2014 targets the thirteen core CODIS loci and twelve core ESS loci, amelo-genin, and the DYS391 locus. There are now five and six dye Fusion system options. The Y-STRs and loci are used in commercial kits for detection of a male contributor or tracing male lineages. The single Y chromosome makes interpretation easier than the diploid autosomal loci. X- and Y-loci can be used in paternity and maternity testing and familial lineage analysis. In 2017, Qiagen introduced the Investigator 24plex GO! Kit and amplifies the CODIS core loci, the ESS markers, plus SE33, D2S1338, D19S433, amelogenin, and DYS391. By March 2018, NDIS contained more than thirteen million offender profiles and more than three million arrestee profiles and 840,000 forensic profiles. The FBI reported CODIS reached 20 million forensic DNA profiles on April 21, 2021 including over 14 million offender profiles, four million arrestee profiles, and over one million forensic profiles in NDIS. Over 558,000 CODIS hits have been reported and CODIS has aided in over 545,000 investigations.

Table 1.2   Core STR Loci for Several Countries and Unions

STR Locus	United States (from 1997)	United States (from 1-1-2017)	United Kingdom	Germany	European Standard Set	Interpol
CSF1PO	X	X				
FGA	X	X	X	X	X	X
TH01	X	X	X	X	X	X
TPOX	X	X				
VWA	X	X	X	X	X	X
D3S1358	X	X	X	X	X	X
D5S818	X	X				
D7S820	X	X				
D8S1179	X	X	X	X	X	X
D13S317	X	X				
D16S539	X	X	X			
D18S51	X	X	X	X	X	X
D21S11	X	X	X	X	X	X
AMEL	X	X	X	X	a	a
D1S1656		X			X	
D2S441		X			X	
D2S1338		X	X			
D10S1248		X			X	
D12S391		X			X	
D19S433		X	X			
D22S1045		X			X	
SE33				X		

a Also included/optional.

CE is regarded as the standard for STR DNA typing. The majority of cases can be solved using CE. It is cost-effective, high-throughput, and offers a fast turnaround. The workflow has been validated and implemented worldwide and is accepted in court. Analysts can obtain a profile for high-profile cases in a few hours, and traditional DNA typing using CE is much lower in cost than emerging methods. There are now eight dye systems that enable the typing of up to thirty STRs simultaneously.

Even with all of the innovations and improvements in the STR typing kits over the last twenty-five years, there are still drawbacks. For example, the major drawback of using electrophoresis-based methods to separate the PCR products in a single run is that the number of loci that can be multiplexed is limited by the size of the amplicons and the number of dye labels the fluorimeter detection system can deconvolute. Another drawback is that fragment sizing does not determine the sequence of the STR repeats, and sequence mutations or variations may aid in the differentiation of samples. More troubling, some experts estimate that 30% of samples do not produce a full STR profile with this method.

Single nucleotide polymorphisms (SNPs) can also be used in human identification. These biallelic marker sites are single-base differences documented to vary between individuals. SNPs are very common throughout the human genome, occurring almost one in every thousand nucleotides. SNPs have been used to predict phenotypic characteristics such as eye color, skin tone, hair color, and biogeographical, ancestral, and behavioral characteristics. An advantage of SNPs is that, owing to their small size, amplifiable DNA fragments containing SNPs are often intact even when STRs sites are not making them ideal for use with degraded DNA. However, they are much less polymorphic than STR loci so more loci are needed to achieve the same discriminating power as the current number of STRs that are used for identification. Additionally, SNPs must be determined using sequencing or sequence-based methods, techniques which permit the determination of each individual DNA base. Sanger sequencing is expensive and slow on the CE instruments that are used for STR sizing. SNaPshot assays have also been developed for SNP determination, but the assays are limited to ten targets per assay and are time consuming to develop and optimize. Figure 1.1 diagrams STR and SNP loci.

(a) rs12913832 eye color SNP on chromosome 15 position 28120472

Brown: ...GCATTAA**A**TGTCA...

Blue: ...GCATTAAGTGTCA...

(b) DYS19 STR (NCBI Accession number X77751) with 12 repeats

TAGG TAGA TAGA TAGA TAGG TAGA TAGA TAGA TAGA TAGA TAGA TAGA TAGA TATA

12 repeats

TAGG TAGA TAGA TAGA TAGG TAGA TAGA TAGA TAGA TAGA TAGA TAGA TAGA TAGA TATA

13 repeats

Figure 1.1 Comparison of a single nucleotide polymorphism and short tandem repeat. (a) rs12913832 is an SNP near the OCA2 gene that has been shown to be linked to blue (G,G) or brown (A,A or A,G) eye color and is used in phenotype prediction assays. (b) DYS19 is an STR (NCBI Accession number X77751) with a variable number of repeats.

1.4 Next Generation Sequencing for Forensic DNA Typing

Massively parallel sequencing (MPS), or next generation sequencing (NGS) as it is widely known, was introduced in the late 1990s by Jonathan Rothberg, and the first commercial instrument was made available in 2005 by 454 Life Sciences (Patrick 2007, Minogue et al. introduced in 2017 by Verogen 2019). NGS methods are faster and higher throughput than Sanger sequencing. While previous DNA typing kits – even the newest kits like GlobalFiler™ and Fusion™ – are limited to twenty-four STR loci, the NGS kits can multiplex more primer pairs as the sequencing method is not size- or dye-limited. NGS-based methods are able to multiplex more STRs and simultaneously type many SNP and X- and Y-haplotype DNA markers in a single assay. For example, the ForenSeq™ kit introduced in 2017 by Verogen types fifty-eight STRs (including twenty-seven autosomal STRs and seven X and twenty-four Y haplotype markers). In addition to typing STR and SNP markers for human identification, SNP markers for determining biogeographical ancestry and phenotype features are determined simultaneously. For example, the ForenSeq™ kit multiplexes a total of 231 loci including ninety-four identity-informative SNPs, fifty-six ancestry-informative SNPs, twenty-two phenotypic-informative SNPs (two ancestry-informative SNPs are used for ancestry and phenotype prediction), and the fifty-eight STRs (Jäger et al. 2017). In 2018, Applied Biosystems introduced the Precision ID NGS kit, and Promega introduced the PowerSeq™ 46GY kit in 2016 that types forty-six loci (van der Gaag et al. 2016). Even a partial NGS profile may lead to more genetic data than was possible using the best CE kits. The NGS amplicons are smaller, which can lead to more genetic data for DNA recovered from degraded and compromised samples. In 2019, the Precision ID and ForenSeq™ data were approved for inclusion in CODIS (Verogen). In NGS, the STR repeat sequences and counts can be compared so that allele subtypes and repeat structure variations can be detected. Isometric heterozygote sequence motifs within STRs and known SNPs in flanking regions can be detected using NGS but not CE. Instead of reporting fluorescence intensity as in CE, the number of reads is outputted in NGS data as shown in Figure 1.2. NGS can be used to type mtDNA and mRNA as well which will be described in Chapters 7 and 9, respectively.

The Scientific Working Groups (SWGs) and Organizational Scientific Area Committees (OSACs) that developed standards for DNA typing updated their documents to include NGS data. The Scientific Working Group on DNA Analysis Methods (SWGDAM) released an addendum to its 2017 interpretation document entitled "SWGDAM Interpretation Guidelines for Autosomal STR Typing by Forensic DNA Testing Laboratories" to address NGS needs in 2019. The US Federal Bureau of Investigation (FBI) published its "Quality Assurance Standards for DNA Databasing Laboratories."

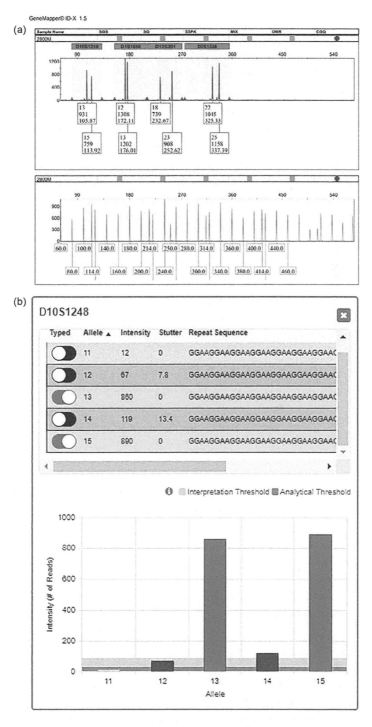

Figure 1.2 Data collected using (a) a modern six-dye kit using CE and (b) NGS. (Courtesy of Adam Klavens.)

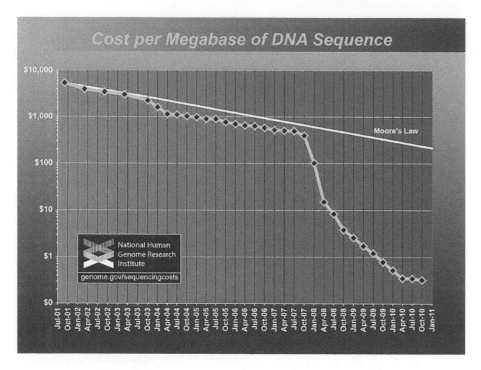

Figure 1.3 Decreasing cost per megabase of DNA sequence over time. (https://search.creativecommons.org/photos/a2a51821-36b0-4432-9a8b-0e86317fca5c.)

1.5 Conclusion

The cost per megabase of DNA sequence has decreased with NGS approaches so that its application for casework is feasible (Figure 1.3). The chapters that follow detail the chemistries used in the various NGS approaches and detail the implementation of NGS tools and analysis methods for forensic DNA typing. The later chapters review NGS applications for mitochondrial DNA typing, body fluids analysis, and bacterial sequencing for forensic applications. NGS offers forensic biologists a myriad new capabilities, but there are still issues to be resolved (Kircher and Kelso 2010, Minogue et al. 2019). The final chapter focuses on the remaining challenges for introducing NGS to more labs and cases.

Questions

1. Define forensic biology.
2. Historically, how did forensic biology develop? Name innovations that changed the field.

3. What precautions must be taken when working with biological evidence to avoid contamination and infection?
4. What is the role of the OSACs in forensic biology?
5. Why have different regions and countries adopted different core STR loci?

References

Anderson, S., Bankier, A.T., Barrell, B.G., de Bruijn, M.H., Coulson, A.R., Drouin, J., Eperon, I.C., Nierlich, D.P., Roe, B.A., Sanger, F., Schreier, P.H., Smith, A.J., Staden, R., and I.G. Young. "Sequence and organization of the human mitochondrial genome." *Nature* 290, no. 9 (April 9, 1981): 457–465. doi:10.1038/290457a0.

Butler, J. *Forensic DNA Typing*, 2nd ed. Burlington, MA: Elsevier Academic Press, 2005.

Edwards, A., Civitello, A., Hammond, H.A., and C.T. Caskey. "DNA typing and genetic mapping with trimeric and tetrameric tandem repeats." *American Journal of Human Genetics* 49, no. 4 (September 30, 1991): 746–756.

Elkins, K.M. *Forensic DNA Biology: A Laboratory Manual*. Waltham, MA: Elsevier Academic Press, 2013.

Erlich, H., Lee, J.S., Petersen, J.W., Bugawan, T., and R. DeMars. "Molecular analysis of HLA class I and class II antigen loss mutants reveals a homozygous deletion of the DR, DQ, and part of the DP region: Implications for class II gene order." *Human Immunology* 16, no. 2 (June 1986): 205–219. doi:10.1016/0198-8859(86)90049-2.

FBI. "Quality Assurance Standards for DNA Databasing Laboratories." Accessed January 23, 2021. https://www.fbi.gov/file-repository/quality-assurance-standards-for-dna-databasing-laboratories.pdf/view.

Franklin, R.E., and R.G. Gosling. "Molecular configuration in sodium thymonucleate." *Nature* 171, no. 4356 (April 25, 1953): 740–741. doi:10.1038/171740a0.

Gill, P., Jeffreys, A., and D. Werrett. "Forensic application of DNA 'fingerprints'." *Nature* 318, no. 6046 (December 12–18, 1985): 577–579. doi:10.1038/318577a0.

Hares, D.R. "Selection and implementation of expanded CODIS core loci in the United States." *Forensic Science International Genetics* 17 (July 1, 2015): 33–34. doi:10.1016/j.fsigen.2015.03.006.

Jäger, A.C., Alvarez, M.L., Davis, C.P., Guzmán, E., Han, Y., Way, L., Walichiewicz, P., Silva, D., Pham, N., Caves, G., Bruand, J., Schlesinger, F., Pond, S.J.K., Varlaro, J., Stephens, K.M., and C.L. Holt. "Developmental validation of the MiSeq FGx forensic genomics system for targeted next generation sequencing in forensic DNA casework and database laboratories." *Forensic Science International: Genetics* 28 (May 2017): 52–70. doi:10.1016/j.fsigen.2017.01.011.

International Human Genome Consortium. "Initial sequencing and analysis of the human genome." *Nature* 409, no. 6922 (February 15, 2001): 860–921. doi:10.1038/35057062.

Kircher, M., and J. Kelso. "High-throughput DNA sequencing-concepts and limitations." *BioEssays* 2, no. 6 (May 18, 2010): 524–536. doi:10.1002/bies.200900181.

Minogue, T.D., Koehler, J.W., Stefan, C.P., and T.A. Conrad. "Next-generation sequencing for biodefense: Biothreat detection, forensics, and the clinic." *Clinical Chemistry* 65, no. 3 (March 1, 2019): 383–392.

Patrick, K.L. "454 life sciences: Illuminating the future of genome sequencing and personalized medicine." *Yale Journal of Biology and Medicine* 80, no. 4 (December 2007): 191–194.

Saiki, R., Scharf, S., Faloona, F., Mullis, K., Horn, G., Erlich, H., and N. Arnheim. "Enzymatic amplification of beta-globin genomic sequences and restriction site analysis for diagnosis of sickle cell anemia." *Science* 230, no. 4732 (December 20, 1985): 1350–1354. doi:10.1126/science.2999980.

Sanger, F., Air, G. M., Barrell, B. G., Brown, N. L., Coulson, A.R., Fiddes, C.A., Hutchison, C. A., Slocombe, P. M., and M. Smith. "Nucleotide sequence of bacteriophage phi X174 DNA." *Nature* 265, no. 5596 (February 24, 1977): 687–695. doi:10.1038/265687a0.

Scientific Working Group on DNA Analysis Methods. "Interpretation Guidelines for Autosomal STR Typing by Forensic DNA Testing Laboratories." Approved January 12, 2017. Accessed January 23, 2021. https://1ecb9588-ea6f-4feb-971a-73265dbf079c.filesusr.com/ugd/4344b0_50e2749756a242528e6285a5bb478f4c.pdf.

Scientific Working Group on DNA Analysis Methods. "Addendum to 'SWGDAM Interpretation Guidelines for Autosomal STR Typing by Forensic DNA Testing Laboratories' to Address Next Generation Sequencing." Approved April 23, 2019. Accessed January 23, 2021. https://1ecb9588-ea6f-4feb-971a-73265dbf079c.filesusr.com/ugd/4344b0_91f2b89538844575a9f51867def7be85.pdf.

van der Gaag, K.J., de Leeuw, R.H., Hoogenboom, J., Patel, J., Storts, D.R., Laros, J., and P. de Knijff. "Massively parallel sequencing of short tandem repeats-Population data and mixture analysis results for the PowerSeq™ system." *Forensic Science International: Genetics* 24 (September 2016): 86–96. doi:10.1016/j.fsigen.2016.05.016.

Venter, C.J., Adams, M.D., Myers, E.W., Li, P.W., Mural, R.J., Sutton, G.G., Smith, H.O., Yandell, M., Evans, C.A., Holt, R.A., and J.D. Gocayne. "The sequence of the human genome." *Science* 291, no. 5507 (February 16, 2001):1304–1351. doi:10.1126/science.1058040.

Verogen. "FBI Approves Verogen's Next-Gen Forensic DNA Technology for National DNA Index System (NDIS)." 2019. Accessed November 7, 2020. https://verogen.com/ndis-approval-of-miseq-fgx/.

Watson, J.D., and F.H.C. Crick. "Molecular structure of nucleic acids: A structure for deoxyribose nucleic acid." *Nature* 171, no. 4356 (April 25, 1953): 737–738. doi:10.1038/171737a0.

Wilkins, M.H.F., Stokes, A.R., and H. R. Wilson. "Molecular structure of deoxypentose nucleic acids." *Nature* 171, no. 4356 (April 25, 1953): 738–740. doi:10.1038/171738a0.

History of Sequencing for Human DNA Typing

2

2.1 Introduction

DNA sequencing has a more than forty-year history which began with Sanger sequencing and has evolved to include newer innovations including SNaPshot assays and next generation sequencing (NGS) methods, including pyrosequencing and massively parallel sequencing. The focus of this chapter is the history and innovations in DNA sequencing and applications to forensic science.

2.2 Common Chemistries Used in Sequencing Applications

Most sequencing chemistries utilize variations of DNA replication found routinely in cells and exploited for *in vitro* sequencing, such as polymerase chain reaction (PCR). In order for replication to occur, a short oligonucleotide called a primer must anneal to single-stranded template DNA. DNA polymerases recognize the double-stranded DNA and bind to this double-stranded complex, and hydrogen bonding allows the nucleotide to be newly added to enter the active site in close proximity to the 5′ end of the priming strand. The 3′ hydroxyl of the terminal nucleotide initiates nucleophilic attack at the 5′ alpha phosphate of the nucleotide to be added resulting in the addition of the nucleotide to the daughter strand. The remaining products in this reaction are pyrophosphate and a proton (Figure 2.1). All of the sequencing strategies that are outlined in this chapter utilize either a fluorescent tag bound to the terminal nucleotide in the elongating daughter strand or the pyrophosphate or proton by-products for the detection of nucleotide incorporation.

2.2.1 Chain Termination Sequencing

Introduced in 1977 (Sanger et al. 1977, Zascavage et al. 2013, Heather and Chain 2016, Bruijns et al. 2018), chain termination sequencing can be used to sequence DNA segments of approximately 800–1000 base pairs, although 500 base pairs or less tend to yield the best results. Prior to sequencing,

DOI: 10.4324/9781003196464-2   13

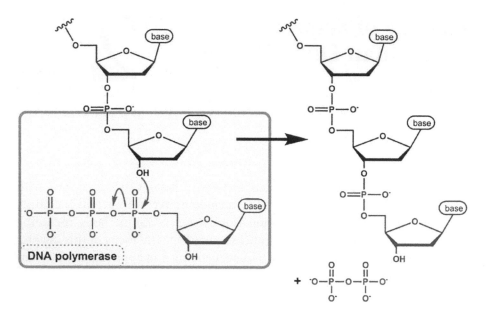

Figure 2.1 DNA synthesis with growing DNA template and pyrophosphate byproduct. (Michał Sobkowski, CC BY 3.0 license.)

extracted DNA is amplified using PCR primers targeting specific loci using a master mix consisting of a high fidelity DNA polymerase, buffer, magnesium, dNTPs, and a 0.1–0.003 X concentration of 2′, 3′-dideoxyribonucleotide triphosphates (ddNTPs). Typically, each of the ddNTPs – ddATP, ddCTP, ddGTP, and ddTTP, which add adenine, cytosine, guanine and thymine nucleotide bases, respectively, – is labeled with a unique fluorescent dye to allow for the identification of the final nucleotide added. The template is extended using the dNTPs and chain-terminating ddNTPs are incorporated at random intervals into the growing nucleotide chain. The sequence is determined following synthesis by separation of the labeled fragments by size using slab or capillary electrophoresis, which has single base pair resolution. This allows for the reading of the sequence of the daughter strand. The process is, therefore, referred to as sequencing by synthesis (SBS) since the sequence is determined by synthesizing the daughter strand (Muzzey et al. 2015) (Figure 2.2).

2.2.2 Pyrosequencing

Pyrosequencing, developed in 1993 as a solid-phase sequencing method, can process lengths up to 300–500 nucleotides, but typically cannot sequence strands as long as can be sequenced by Sanger sequencing. The extracted DNA is digested to ~100 bp fragments, denatured to form single-stranded

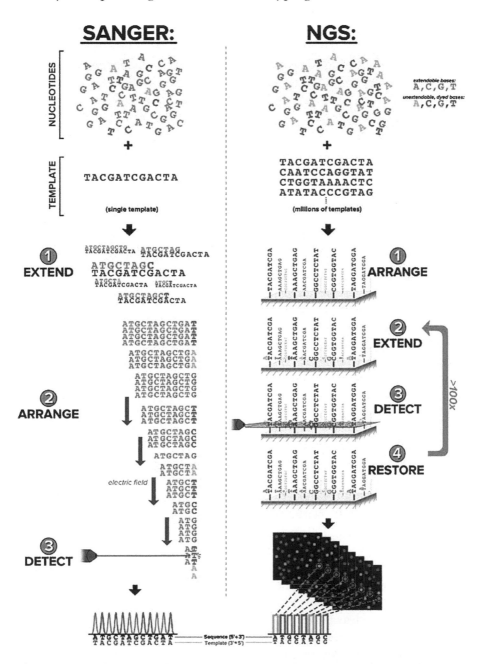

Figure 2.2 Comparison of Sanger sequencing and NGS processes. (Dale Muzzey, Eric A. Evans, Caroline Lieber, CC BY 4.0 license. https://www.ncbi.nlm.nih.gov/pmc/articles/PMC4633438/figure/Fig1/.)

DNA (ssDNA), and attached to the surface of beads using adaptors or linkers; or the region of interest can be amplified using PCR, and the single-stranded products are then annealed to the beads. In the pyrosequencing reaction, the dNTPs are dispensed in a user-defined order and those incorporated result in the release of pyrophosphate (PPi). In an enzyme cascade, sulfurylase uses the released pyrophosphate in the presence of adenosine-5′-phosphosulfate (APS) to generate ATP in equimolar concentration. Luciferase in the presence of ATP converts luciferin to oxyluciferin to generate light proportional to the number of nucleotides added in that particular addition step. Apyrase is used to degrade unincorporated nucleotides and allow the process to reset prior to the next dispensation of nucleotides (Nyren et al. 1993, Ronaghi et al. 1996, Bruijns et al. 2018).

In order to perform *de novo* synthesis, repeated rounds of dispensing all four nucleotides must be performed, making pyrosequencing relatively inefficient for *de novo* sequencing long stretches of DNA. The dispensation of nucleotides in pyrosequencing can be defined by the user in order to maximize efficiency. Pyrosequencing can be a cost-effective way to confirm sequence or check for site mutations in a particular DNA fragment. It is also widely used in the determination of methylation status of genes. In both of these cases, the dispensation is optimized based on the expected sequence of the amplicon (Figure 2.3).

2.2.3 Sequencing by Ligation

Another approach to sequencing involves sequencing by ligation (Figure 2.4). Short probes, seven to nine nucleotides in length, are annealed to the template strand of DNA. The probes are designed in such a way that there are 16 different probes labeled with four different fluorescent tags. The 3′ end of each probe has two annealing nucleotides followed by three degenerate nucleotides, a cleavage site, three additional degenerate nucleotides, and one of four fluorescent dyes that is associated with the two 3′ nucleotides. Upon annealing, the probes are ligated to the daughter strand, unannealed nucleotides are washed away, and the fluorescent tag is read. The reset to begin the next cycle is completed by cleavage of the probe between nucleotides five and six. This allows for the determination of the first two nucleotides and then has a gap of three unknown nucleotides prior to the next cycle of annealing. Usually, five to seven cycles of annealing are performed per sequencing primer. In order to obtain a complete sequence, multiple sequencing primers are utilized in additional rounds of sequencing. Each of the new sequencing primers is one nucleotide shorter in length and therefore interrogates the sequence offset by a single nucleotide. Obtaining the full sequence of twenty-five to thirty-five nucleotides in length requires five rounds of sequencing of five to seven cycles.

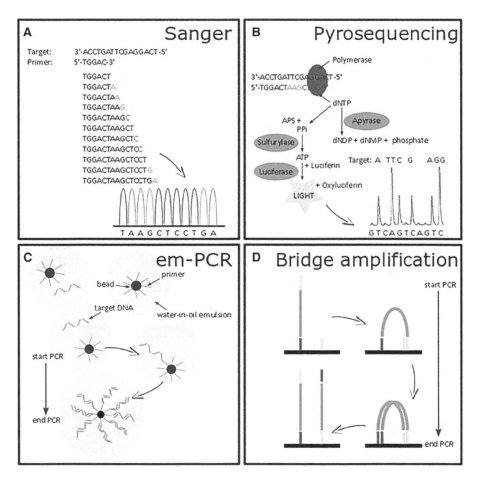

Figure 2.3 Overview of several DNA sequencing techniques with the principle of (a) Sanger sequencing, (b) pyrosequencing (e.g. 454), (c) em-PCR (e.g., 454,
SOLiD® and Ion Torrent™), and (d) bridge amplification/cluster PCR (e.g., Solexa).
(Brigitte Bruijns, Roald Tiggelaar, and Han Gardeniers, CC BY NC ND 4.0 license.
https://www.ncbi.nlm.nih.gov/pmc/articles/PMC6282972/figure/elps6707-
fig-0002/?report=objectonly.)

2.3 Detection Techniques

In order to detect nucleotide incorporation and determine the identity of the
added base, one of several detection methods is used.

2.3.1 Fluorescence

Fluorescence dyes are commonly used for detection of nucleotides. Dyes
are covalently attached to bases, which sterically hinders the reaction from

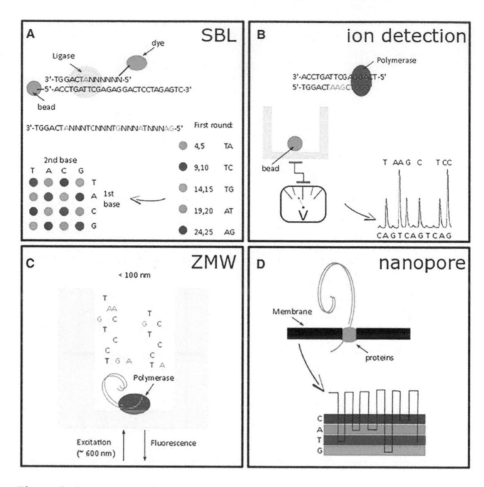

Figure 2.4 Overview of several DNA sequencing techniques with the principle of (a) sequencing by ligation (SBL, e.g., SOLiD®), (b) ion detection (e.g. Ion Torrent™), (c) zero-mode waveguides (ZMWs, e.g., PacBio®), and (d) nanopores (e.g., Oxford Nanopore). (Brigitte Bruijns, Roald Tiggelaar, and Han Gardeniers, CC BY NC ND 4.0 license. https://www.ncbi.nlm.nih.gov/pmc/articles/PMC6282972/figure/elps6707-fig-0003/?report=objectonly.)

proceeding. The dyes are excited using ~500 nm wavelength laser causing them to emit light at emission maxima from 520 to 625 nm depending on the dye (Butler 2005, Elkins 2013). Each dye is associated with an individual nucleotide. All of the nucleotides are added in a single reaction, and the identification of the nucleotide is directly determined by the wavelength maxima of the emission spectra. The detection requires the use of filters to detect, the amount of light emitted by the fluorophores at each wavelength at each shutter opening which is recorded using a charge-coupled device (CCD) camera.

2.3.2 Pyrosequencing

The production of pyrophosphate by the DNA polymerase in the DNA synthesis reaction is converted to blue-green light, through the enzymatic cascade shown in Figure 2.2. Detection requires the use of a CCD camera. The camera is able to detect very low levels of light and record the amount of light recorded at each shutter opening. The light produced is proportional to the number of same type of nucleotides added at a time. Since the dispensation order is determined prior to the reaction proceeding, the presence of light indicates that the dispensed nucleotide was added; if no light is produced, no nucleotide addition is indicated.

2.3.3 Ion Detection

Unlike the previously described optical detection systems, ion detection does not require modification of the nucleotide or the presence of dyes and uses only standard nucleotides. The production of an H^+ by the DNA synthesis reaction can be detected by an ion-sensitive field-effect transistor. The reaction occurs in a well above an ion-sensitive layer which is above the ion-sensitive field-effect transistor (ISFET) detector. As in pyrosequencing, the nucleotides are dispensed in a specific order, and the signal associated with the addition relates to the number of nucleotides added at a time, thereby indicating the length of a homopolymeric stretch.

2.4 Sequencing Platforms

Each sequencing platform requires a chemistry method, detection method, and an interpretation method. Examples of how these modules are combined to produce different types of sequencing platforms are described below.

2.4.1 First-Generation Sequencing Techniques

First-generation technologies have the capability to sequence only one sample at a time. The individual sample is interrogated with a particular chemistry. Following detection, interpretation is usually simple, either by determination of the order of a color sequence, or by a conformation of the addition of a particular nucleotide in the dispensation.

2.4.1.1 Sanger Sequencing

Sanger sequencing was introduced in 1977 (Sanger et al. 1977). Prior to sequencing, the extracted DNA is often amplified using PCR primers targeting specific loci or degenerate primers that randomly bind to locations in the genome depending on the type of sequencing desired. Targeted sequencing to determine a particular gene sequence uses specific primers. Degenerate primers are added to amplify sequences noted to have variants in the primer binding region. Each of the ddNTPs – ddATP, ddCTP, ddGTP, and ddTTP – is labeled with a unique fluorescent dye. The template is extended using the dNTPs, and chain-terminating ddNTPs are incorporated into positions, at random, into the growing nucleotide chain. The sequence is determined following synthesis, therefore sometimes referred to as sequencing by synthesis. Following PCR, the amplicons are separated by size using electrophoresis, and the identity of the bases in sequence is determined by the location of each base by the fluorescent color detected. Initially, the sequence was determined using large slab gels and electrophoresis. In 1995, Applied Biosystems introduced the first capillary electrophoresis instrument, the Applied Biosystems 310 Genetic Analyzer, a single capillary instrument with four-dye fluorescence detection. For comparison, the modern Applied Biosystems 3500 Genetic Analyzer is an eight-capillary instrument equipped with six-dye detection. CE provides single-nucleotide spatial resolution, good spectra resolution for dye separation, and precision in DNA sizing. The CE separation takes approximately an hour. The CE output is termed an electropherogram. Revolutionary for its time, Sanger sequencing enabled researchers to sequence the first genome (that of bacteriophage phiX174) and those of many more species after it including the 3.2 billion base pair human genome as reported in a pair of *Science* and *Nature* papers in 2001 (International Human Genome Consortium 2001, Venter et al. 2001). During this time, Craig Venter pioneered the whole genome shotgun (WGS) sequencing approach. Following sequencing, the target is analyzed by comparison to a known (K) standard or reference sample or a sequenced human genome standard. The price of Sanger sequencing at core facilities has fallen to approximately $4–$10 per sample. Although capillary electrophoresis and Sanger sequencing yield discrete single-base resolution and identification (Figure 2.5), the DNA typing method is relatively slow and limited in read length.

2.4.1.2 SNaPShot Sequencing

SNaPshot is a chain termination minisequencing or a primer extension assay based on a Sanger sequencing approach (Butler 2005). Primers are designed to complement the region directly upstream of an SNP so that the SNP base is added by the DNA polymerase in the DNA sequencing reaction. The ddNTP bases used in the PCR reaction are dye-labeled to facilitate fluorescence detection. Poly(T) mobility tails can be added to the primers to aid in size separation during electrophoresis. For example, primer one may have a five base poly(T) 5'-tail, primer two could have a ten base poly(T) tail, primer three could have a fifteen base poly(T) tail, and so on. The primers can be multiplexed so that multiple SNPs can be determined simultaneously. The genomic DNA is first amplified using PCR. Exonuclease I is used to digest unincorporated single-stranded primers. The Applied Biosystems SNaPshot kit contains the ddNTPs each with

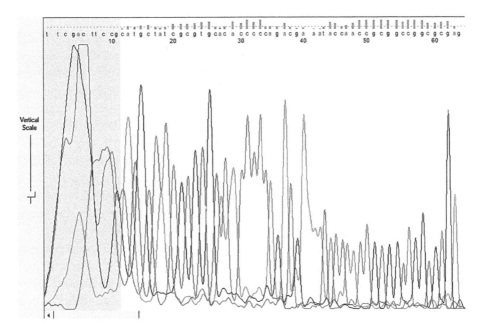

Figure 2.5 A sample Sanger sequence read.

a unique dye label (A-green, G-blue, C-yellow, T-red), buffer, DNA polymerase, and an internal standard labeled with a fifth dye in the master mix. In the second round of PCR, the SNaPshot master mix is used to perform the primer extension. As ddNTPs rather than a mixture of ddNTPs and dNTPs are used in the sequencing reaction, incorporation of the base immediately stops chain extension as no further bases can be added without the 3′-hydroxyl. Following PCR, shrimp alkaline phosphatase (SAP) is used to digest unincorporated ddNTPs to reduce dye artifacts. Following PCR, the amplicons are resolved by size using CE, using POP-4 (or POP-6 or POP-7) for separation, and using the GS120 LIZ size standard for sizing. GeneMapper*ID* is used to analyze the results on Applied Biosystems instruments. Unlike using CE for STR DNA typing, there is no stutter artifact in SNaPshot DNA typing.

2.4.1.3 Pyrosequencing

Pyrosequencing is a SBS method that is widely used to detect and quantify CpG and other types of methylation sites and SNP heteroplasmy. CpG methylation occurs in the 5′ upstream untranslated region of a gene, controlling gene expression. It is observed through bisulfite treatment, which converts unmethylated cytosines to uracils through a deamidation reaction catalyzed by sodium bisulfite. Methylated cytosines remain unchanged. SNP heteroplasmy is the presence of two different SNP variant in a single individual. This can occur in the same cell or in within a population of cells. Pyrosequencing is ideally suited to determining the ratio of two different nucleotides at a particular site.

Pyrosequencing can be used to determine the identity of one or more SNPs, ideally in a short target region of twenty to thirty nucleotides, although some sources suggest that it can be used for lengths up to 300–500 nucleotides. While it can target more than one base per primer unlike the SNaPshot method, pyrosequencing cannot sequence strands as long as can be sequenced by Sanger sequencing. The extracted DNA is digested to ~100 bp fragments and denatured to form single-stranded DNA (ssDNA) and attached to the surface of beads using adaptors or linkers. Samples are prepared by emulsion PCR prior to pyrosequencing using biotinylated PCR primers as follows. The DNA is attached to microscopic beads. The beads are each dispersed into a droplet in a water–oil emulsion where the PCR reaction occurs and amplifies the target. In the pyrosequencing reaction, the dNTPs are added in a specific order and those incorporated result in the release of pyrophosphate. In an enzyme cascade, sulfurylase uses the released pyrophosphate to form ATP in the presence of adenosine monophosphate (AMP). Luciferase uses the ATP to convert luciferin to oxyluciferin in the presence of adenosine-5′-phosphosulfate and generates light. The light is detected by the instrument and indicates that the base released by the instrument was incorporated into the growing chain. Apyrase degrades nucleotides not added to the chain. The process works well even for tetramer homopolymeric sections of DNA. Pyrosequencing takes approximately two hours and costs approximately $10 per sample. Pyrosequencing was purchased by Qiagen in 2008 and licensed by 454 Life Sciences, which developed an array-based pyrosequencing instrument (discontinued in 2013). Another pyrosequencer was the GS FLX Titanium. Qiagen's latest PyroMark Q48 pyrosequencer (Figure 2.6), the company's newest pyrosequencing instrument, can sequence forty-eight samples simultaneously.

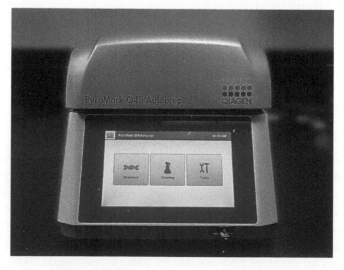

Figure 2.6 Qiagen Q48 pyrosequencing instrument.

2.5 Massively Parallel Sequencing

Massively parallel sequencing (MPS) is a type of NGS and one of the newest sequencing methods. In MPS, millions of sequencing reads are detected on the same chip or flow cell in parallel. NGS, or second-generation sequencing, instruments emerged in the mid- to late 1990s and became available for commercial purchase in 2005 with the $1000 genome becoming a reality in 2014 (van Dijk et al. 2014, Heather and Chain 2016). An early application of this chemistry was described in 2008 (Bentley et al. 2008). The earliest applications were in the clinical life sciences for disease diagnosis. Ancestry applications have emerged more recently. Several manufacturers, including Illumina, Life Technologies, Complete Genomics, Helicos Biosciences, Oxford, and Pacific Biosciences, developed different chemistry and approaches for sequencing. Life Technologies has been acquired by Applied Biosystems which was acquired by ThermoFisher.

2.5.1 Reversible Chain Termination MPS Platforms

The Illumina sequencers include the iSeq, NanoSeq, MiSeq, HiSeq, and the Genome Analyzer IIX. The Illumina technology employs clonal bridge amplification to produce a high concentration of clonal DNA that is covalently attached to the flow cell. It can perform single and paired ends reads and uses reversible dye sequencing by synthesis chemistry (Figure 2.2). Using a CCD camera and fluorescence detection, individual nucleotide addition at each cluster is evaluated at every cycle. Each end read can be up to 150 bases. They vary in how many flow cells and the number of bases that can be read in a run. All of the Illumina instruments except the HiSeq process one flow cell per run. The HiSeq can process two flow cells in parallel and generate 600 billion bases in one run for the highest instrument cost but the lowest cost per base of the Illumina instrument. The Verogen MiSeq FGx is built on the Illumina MiSeq platform; the instrument has a small footprint and is shown in Figure 2.7.

2.5.2 Ion Detection Platforms

The Applied Biosystems sequencers include the Ion Torrent, Ion S5, Ion Proton, and Ion PGM systems and employ an approach introduced in 2010 in which a pH change is converted to a base call. As with the pyrosequencing technologies, the ion detection platforms detect the same by-product of synthesis for each addition, in this case a proton; therefore, the order of nucleotide dispensations must be known. The voltage change is recorded when a proton ion is released upon addition of a nucleotide base in a process

Figure 2.7 Verogen MiSeq FGx instrument.

called semiconductor sequencing. This type of sequencing does not require the alteration of nucleotides or the addition of an enzymatic cascade for detection.

2.5.3 Sequencing by Ligation Platforms

Applied Biosystems SOLiD instrument employs a ligation and detection approach introduced in 2005 with a commercial release in 2007 (Kircher and Kelso 2010). The sequencing is performed by DNA ligase instead of DNA polymerase. The extracted template DNA is fragmented using restriction enzymes, and P1 and P2 adaptors are added by PCR. The fragments are annealed to clonal emulsion beads via the universal P1 adaptor.

One- or two-base encoded fluorescently labeled probes are attached to a primer hybridized to the complementary target and are joined by DNA ligase. Upon ligation, a fluorophore is cleaved off and the identity of the ligated probe is determined using fluorescence emission. Four sequencing primers "n" to "n-4" by each base so the process is termed oligonucleotide eight-mer chained ligation chemistry. Non-ligated probes are washed away. The process continues by cleaving the probe to regenerate the 5'-OH group. The maximum read length is a relatively short twenty to forty-five bases; therefore, overlapping regions sequenced must be ordered *in silico*. The sequence is constructed by overlapping the dinucleotides. The process has a very low single base error rate and is quite inexpensive per megabase.

2.5.4 Single Base Extension Platforms

Similar to the technology used for the SNaPshot assay, single base extension platforms, such as the Illumina iScan, use a combination of primers specific to the SNPs of interest and fluorescently tagged dNTPs. The template DNA undergoes whole genome amplification, followed by fragmentation prior to annealing the template DNA to oligonucleotide primers for SNPs bound to beads, which are specifically located on a chip. The primed template undergoes one round of extension to label the SNP with a fluorescent nucleotide which indicates the allele at the site.

2.5.5 Third-Generation Platforms

Third-generation platforms use often use nanofluidics and are capable of detecting sequences of DNA or RNA from a single cell in real time. These platforms often use the same sequencing chemistries described for the second (next) generation platforms; however, the scaling is much smaller.

The Oxford Nanopore instruments including the MinION conduct real-time sequencing using nanopore technology in a portable, USB-sized device that emerged in 2014. This instrument uses ion detection chemistry in a reduced size platform. Other instruments include the SmidgeION, flongle, GridION, and PromethION. The read length extends to 2000 bases and up to 30 GB per run. Interpretation includes demultiplexing barcodes assigned to target regions as in the larger second-generation instruments.

The Helicos BioSciences Heliscope also uses reversible dye terminator chemistry with a solid-phase, PCR free, primer extension approach. The maximum read length is thirty-five bases.

Complete Genomics employs a process termed unchain ligation using nine-mer oligonucleotides and rolling circle amplification.

The Pacific Biosciences Single-Molecule Real-Time (SMRT) instrument chemistry uses single-molecule chemistry with a polymerase attached to a solid support which extends primed templates using phospholinked fluorescent nucleotides. It does not require PCR. The Sequel II model can sequence up to four million reads and up to 500 GB raw read data.

A summary of NGS read length, run time, and per-base cost for the sequencing approaches and platforms is tabulated in Table 2.1. Advantages and disadvantages of the various sequencing approaches are tabulated in Table 2.2 (Berglund et al. 2011).

2.6 NGS Instruments Adopted for Forensic Science

The availability of commercial kits to process forensic samples is an important factor for labs deciding to add NGS to their repertoire and adopt a

Table 2.1 NGS Read Length, Run Time, and Per Base Cost

Sequencing Approach	Method	Read Length (Typical) (bp)	Run Time	Cost (Per Mb, Approx. USD)	Instrument Cost (Approx. USD)
Sanger	SBS (fluorescent ddNTPs)	1000 (20–450)	45/capillary (set)	$500/Mb	$90,000/ABI 3500/8 capillary
Pyrosequencing	SBS (luciferase)	140 (35)	1 minute/bp	$100/Mb	$80,000
Roche/454	Emulsion PCR SBS (pyrosequencing)	250	9 hours	$20/Mb	$500,000
MiSeq	SBS (reversible terminators)	150×2 or 300×2	27 hours/96 samples	$0.50/Mb	$97,000
Ion Torrent PGM	SBS (H⁺ detection)	>100	2.5–4 hours	$0.63/Mb	$80,000
SOLiD	Emulsion PCR (ligation)	75 (8)	14 days	$0.50/Mb	$591,000
SMRT	Single molecule sequencing, SBS	>2000	20–30 hours, real-time	$2/Mb	$695,000
Solexa	Bridge PCR (reversible terminators)	36	1–10 days	$2/Mb	$430,000
HeliScope	Single molecule (asynchronous extension)	35	8 days	<$0.50/Mb	$1,350,000
Polonator	Emulsion PCR (ligation)	13	4 days	$1/Mb	$170,000

SBS, sequencing by synthesis.

Table 2.2   Advantages and Disadvantages of Various Sequencing Approaches

Sequencing Approach	Advantages	Disadvantages
Sanger	Inexpensive for short reads, accessible, short run time per sample, excellent for *de novo* sequencing	Overall high cost per Mb, targeted sequencing, difficulty with homopolymeric stretches
Pyrosequencing	Inexpensive for short reads, short run time per sample, outputs percent methylation, can sequence homopolymeric stretches, excellent for SNPs	Most expensive per Mb than Sanger, short reads, slow for de novo sequencing
Roche/454	At introduction greatly reduced cost per Mb, paired-end sequencing for dual confirmation, long reads	High initial cost, high cost for analysis, must multiplex to be cost-effective
MiSeq	Inexpensive per Mb, great option for low-medium throughput lab, paired-end sequencing for dual confirmation, flexible, add indexes using standard PCR, portable sister instruments (e.g., iSeq), easy to use, robust, compatible with other manufacturers products	Must multiplex to be cost-effective
Ion Torrent	Inexpensive per Mb, great option for low-medium throughput lab, paired-end sequencing for dual confirmation, short run time, low run time, robust, compatible with other manufacturers products	Must multiplex to be cost-effective
SOLiD	Inherent error correction Inexpensive per Mb	High initial cost, must multiplex to be cost-effective, time-consuming data processing, short reads
SMRT	Long reads, short run time	High initial cost, must multiplex to be cost-effective, high error rate
Solexa	Can generate over a billion bases per run, highly accurate	High initial cost, must multiplex to be cost-effective, short read length
HeliScope	Sequences RNA, inexpensive per Mb	High initial cost, must multiplex to be cost-effective, high error rate, short read length
Polonator	Inexpensive per Mb	Long post-sequencing assembly time, short read length

specific instrument or approach. A review of NGS and its forensic genetics applications was published in 2015 (Børsting and Morling 2015), although new kits and capabilities continue to be introduced. Time constraints and pressure to process casework as quickly as possible can limit research and development in forensic labs. Furthermore, after the sequencing data is collected, lab must have tools to analyze it and facilitate reporting. To date, the Illumina MiSeq/Verogen MiSeq FGx (Caratti et al. 2015) and Applied Biosystems/ThermoFisher series of instruments including the Ion Proton, Ion PGM, Ion S5, and Ion Torrent have been adopted for forensic use. In 2017, the ForenSeq DNA Signature Prep kits were made available for human identity testing using the Illumina MiSeq NGS instrument; the Illumina platform has also been adopted for sequencing libraries with the Promega PowerSeq 46GY kit. The Ion series instruments have been adopted for sequencing libraries prepared using the Precision ID GlobalFiler NGS STR Panel v2, HID-Ion Ampliseq™ Ancestry Panel, HID-Ion Ampliseq™ Identity Panel, QIAseq Investigator Panels, and GenPlex™ HID kits. Forensic applications of these and emerging tools are the focus of the rest of this book.

Questions

1. Which sequencing method(s) is (are) the most cost-effective for determining the sequence of a short read? Explain your answer.
2. Which sequencing method(s) is (are) the most cost-effective for determining the sequence at an SNP site? Explain your answer.
3. Which sequencing method(s) is (are) the most cost-effective for determining the sequence of tens of loci or a genome? Explain your answer.
4. Which sequencing method should not be used for STR DNA typing? Explain your answer.
5. Compare and contrast the sequencing methods in terms of read length, processing time, and per base cost.

References

Bentley, D.R., Balasubramanian, S., Swerdlow, H.P., Smith, G.P., Milton, J., and C.G. Brown. "Accurate whole human genome sequencing using reversible terminator chemistry." *Nature* 456, no. 7218 (November 6, 2008): 53–59. doi:10.1038/nature07517.

Berglund, E.C., Kiialainen, A., and A.-C. Syvänen. "Next-generation sequencing technologies and applications for human genetic history and forensics." *Investigative Genetics* 2 (November 24, 2011): 23. doi:0.1186/2041-2223-2-23.

Børsting, C., and N. Morling. "Next generation sequencing and its applications in forensic genetics." *Forensic Science International Genetics* 18 (September 2015): 78–89. doi:10.1016/j.fsigen.2015.02.002.

Bruijns, B., Tiggelaar, R., and H. Gardeniers. "Massively parallel sequencing techniques for forensics: A review." *Electrophoresis* 39, no. 21 (August 13, 2018): 2641–2654. doi:10.1002/elps.201800082.

Butler, J. *Forensic DNA Typing*, 2nd ed. Burlington, MA: Elsevier Academic Press, 2005.

Caratti, S., Turrina, S., Ferrian, M., Cosentino, E., and D. De Leo. "MiSeq FGx sequencing system: A new platform for forensic genetics." *Forensic Science International: Genetics Supplement Series* 5 (August 2015): e98–e100. doi:10.1016/j.fsigss.2015.09.040.

Elkins, K.M. *Forensic DNA Biology: A Laboratory Manual*. Waltham, MA: Elsevier Academic Press, 2013.

Heather, J.M., and B. Chain. "The sequence of sequencers: The history of sequencing DNA." *Genomics* 107, no. 1 (January 2016): 1–8. doi:10.1016/j.ygeno.2015.11.003.

International Human Genome Consortium. "Initial sequencing and analysis of the human genome." *Nature* 409, no. 6922 (February 15, 2001): 860–921. doi:10.1038/35057062.

Kircher, M., and J. Kelso. "High-throughput DNA sequencing – Concepts and limitations." *BioEssays* 32, no. 6 (June 2010): 524–536. doi:10.1002/bies.200900181.

Muzzey, D., Evans, E.A., and C. Lieber. "Understanding the basics of NGS: From mechanism to variant calling." *Genetic Counseling and Clinical Testing* 3 (December 2015): 158–165. doi:10.1007/s40142-015-0076-8.

Nyren, P., Petersson, B., and M. Uhlen. "Solid phase DNA minisequencing by an enzymatic luminometric inorganic pyrophosphate detection assay." *Analytical Biochemistry* 208, no. 1 (January 1993): 171–175. doi:10.1006/abio.1993.1024.

Ronaghi, M., Karamohamed, S., Pettersson, B., Uhlén, M., and P. Nyrén. "Real-time DNA sequencing using detection of pyrophosphate release." *Analytical Biochemistry* 242, no. 1 (November 1, 1996): 84–89. doi:10.1006/abio.1996.0432.

Sanger, F., Air, G. M., Barrell, B. G., Brown, N. L., Coulson, A.R., Fiddes, C.A., Hutchison, C. A., Slocombe, P. M., and M. Smith. "Nucleotide sequence of bacteriophage phi X174 DNA." *Nature* 265, no. 5596 (February 24, 1977): 687–695. doi:10.1038/265687a0.

van Dijk, E.L., Auger, H., Jaszczyszyn, Y., and C. Thermes. "Ten years of next-generation sequencing technology." *Trends in Genetics* 30, no. 9 (September 2014): 418–426. doi:10.1016/j.tig.2014.07.001.

Venter, C.J., Adams, M.D., Myers, E.W., Li, P.W., Mural, R.J., Sutton, G.G., Smith, H.O., Yandell, M., Evans, C.A., Holt, R.A., and J.D. Gocayne. "The sequence of the human genome." *Science* 291, no. 5507 (February 16, 2001): 1304–1351. doi:10.1126/science.1058040.

Zascavage, R.R., Shewale, S.J., and J.V. Planz. "Deep sequencing technologies and potential applications in forensic DNA testing." *Forensic Science Review* 25, no. 1–2 (2013): 79–105.

3

Sample Preparation, Standards, and Library Preparation for Next Generation Sequencing

3.1 Overview of the NGS Sample Preparation Process

In order to analyze samples using next generation sequencing (NGS), several steps must be performed. An overview of the steps in the NGS sample preparation process is shown in Figure 3.1. The process begins with sample handling and processing followed by DNA extraction and DNA quantitation. Upon extracting and ascertaining the quantity and condition of the DNA samples, a process termed library preparation is begun. Following library preparation, samples undergo purification and normalization steps prior to multiplexing and denaturation. Sequencing can then be performed; it is the focus of Chapter 4.

3.2 Sample Handling and Processing

The samples chosen for DNA typing using NGS may be old, degraded, or extremely limited in quantity or they may be pristine, fresh body fluid samples. Because NGS is known to have a greater sensitivity than multiplex PCR kits analyzed using the CE, it is extremely important to reduce and avoid potential sources of contamination. Surfaces and non-autoclavable equipment should be cleaned to reduce or eliminate residual DNA using at least two of the following approaches: soap and water, UV light irradiation, 10% bleach, 70% ethanol, Cidex, LookOut® DNA Erase Spray, and/or DNAZap. Crushing and pulverizing bone and teeth should be performed in a hood to reduce dust carry over between samples and the work area. Stainless steel equipment including freezer mill parts and die presses used to crush and pulverize bone and teeth should be cleaned and autoclaved between each use. When working with reagents and samples, DNase and RNase-free tubes and aerosol filter tips should be used. Samples should be processed sequentially in a preparation space separate from the amplification area. Analysts should wear N95 protective masks to reduce dust inhalation when sanding and crushing skeletal remains samples. Other personal protective equipment (PPE) including goggles, gloves, hair nets, masks, lab coats, and closed-toed shoes should always be worn when processing samples in the DNA lab.

DOI: 10.4324/9781003196464-3   31

Figure 3.1 Overview of steps in the NGS sample preparation process.

3.3 DNA Extraction

There are several methods that can be used to extract DNA from cellular material. While the organic phenol-chloroform-isoamyl alcohol (PCI or PCIA) (Butler 2005) and the inorganic metal-chelating Chelex-100 resin (Walsh et al. 1999) were widely used for forensic DNA extraction in the past, silica-based methods now dominate (Hoff-Olsen et al. 1999, Castella et al. 2006, Brevnov et al. 2009, Elkins et al. 2013, Eychner et al. 2017). While the PCIA method remains the gold standard for some applications because it yields the highest quantity of DNA from samples and the double-stranded structure remains intact, care must be taken to remove all organics and phenol as the phenol hydroxyl group will attack the phosphodiester bonds in DNA strands and lead to fragmented, degraded DNA. Furthermore, the PCIA method is not ideal for all samples; for example, it digests the gum in chewing gum (Eychner et al. 2017). The Chelex-100 method is an inexpensive method that requires little hands-on time, although it, like the PCIA method, requires an overnight incubation step that must be accommodated in the lab's standard operating procedure (SOP).

Most of the modern DNA extraction methods incorporated into commercial kits are silica-based methods. Examples are the Qiagen QIAamp DNA Investigator and EZ1 DNA Investigator kits which employ magnetic bead suspensions. Other approaches, such as the Applied Biosystems PrepFiler and the Promega DNA IQ kits, employ magnetic beads and magnetite-modified silicon dioxide magnetic beads, respectfully. These methods are rapid, can be automated, and yield the most highly reproducible DNA yields (Eychner, et al. 2017). Robotic tools include the Beckman Coulter Biomek liquid handling platform, Tecan HID EVOlution™, Promega Maxwell® and Maxprep™ Instruments, and Qiagen QIAcube HT®, QIAcube Connect, and EZ1 instruments. For example, the EZ1 DNA Investigator Kit is used with the Qiagen EZ1 robot. Samples collected on FTA cards or similar fibrous material containing preservatives can be extracted using one of the above kits such as the EZ1 DNA Investigator kit or methods or added directly to the PCR sample tube in library preparation (Kampmann et al. 2016). As modern human DNA typing kits typically require 1 ng or less of input DNA, any of the extraction methods can typically recover sufficient DNA from blood or buccal samples.

Many DNA extraction methods have been explored for extracting DNA from bone and teeth samples. Since the DNA is protected in the bone cells and marrow and tooth pulp, the samples first undergo cleaning, decalcification, cutting, crushing, and drilling steps. In a recent study, the PrepFiler® BTA method performed better than a PCI-silica-based method for extracting DNA from bone (Hasap et al. 2019). In our lab, a modified procedure employing the Qiagen EZ1 DNA Investigator kit silica-based method and EZ1 BioRobot has performed the best for extracting DNA from modern and historic bone for use in DNA typing analysis (Dukes et al. 2012, Klavens et al. 2020). However, in yet another recent study, some crude bone lysates were shown to inhibit DNA purification when using paramagnetic silica beads in the DNA IQ™ Casework Pro Kit and the Maxwell® sixteen due to the filter clogging and were enhanced when phenol was used to treat the lysates (Desmyter et al. 2017). Edson (2019) reported that the best DNA recovery from postcranial osseous human remains of service members lost in World War II, the Korean War, and South-East Asia was obtained using a complete demineralization method with organic extraction when four DNA extraction protocols were evaluated. Zupanič Pajnič et al. (2020) quantified human remains from a massacre of a Slovenian family during World War II using the PowerQuant kit and were able to produce DNA profiles from a molar and five femurs. Zeng et al. (2019) tested the DNA IQ™, DNA Investigator, and PrepFiler® BTA kit as well as the organic extraction method for hair and blood and two total demineralization protocols for bone; the team found all of the DNA extraction methods to be efficient and compatible with the Precision ID and ForenSeq kits. Carrasco et al. (2020) reported upon a specialized extraction method for degraded blood and dental remains samples. Sidstedt et al. (2020) evaluated PCR inhibition and MPS applications.

DNA recovery of trace materials including hair, fingernails and fingerprints, and other touch DNA has also been studied. Naue et al. (2020) found that rubbing a wet swab on a hair led to the recovery of DNA from the contributor of the surface material, and the individual who was the source of the hair and cleaning the hair is essential to obtain a single source hair profile. Preuner et al. (2014) found that the PrepFiler Forensic DNA Extraction kit led to high-quality DNA from fingernail clippings. Tasker et al. (2017) extracted DNA from post-blast improvised explosive device (IED) pipe bombs. England et al. (2020) tested the ForenSeq kit with DNA extracted from laser microdissected cells. Eychner et al. (2017) tested five DNA extraction methods (PCIA, QIAamp DNA Investigator, DNA IQ, Chelex-100, and PrepFiler) for recovering DNA from chewing gum and saliva aliquoted on swabs; the QIAamp method performed the best overall. For low-quantity samples such as chewing gum, touch DNA and human remains, ethanol precipitation, concentrator devices, and reduced elution volumes can be employed to concentrate the recovered DNA (although larger elution volumes will maximize the overall yield) (Moore 1998, Eychner et al. 2017).

3.4 DNA Quantitation

There are several DNA quantitation options for evaluating the quantity of DNA recovered to determine how much of the extract to input for NGS typing. The quantitation methods include human-specific and non-specific methods. Spectroscopic methods for quantifying DNA include UV-Vis and fluorescence spectroscopy. UV-Vis Spectroscopic methods can be used to estimate total DNA in a sample, but they are not human-specific (Elkins 2013). The NanoDrop (UV-Vis spectroscopy) and Qubit (fluorescence spectroscopy) spectrophotometers are widely used for rapid DNA quantification as they require as little as one microliter of sample.

Real-time PCR methods coupled with fluorscence detection can be used to determine the quantity of human DNA and also determine its amplifiability (Horsman et al. 2006). Both "home brew" and commercial assays can be employed. A real-time PCR assay targeting the TPOX locus quantitates total human and male genomic DNA (Horsman et al. 2006). Several commercial kits are available to determine the quantity of human DNA in a sample and simultaneously determine the presence of human and male genomic DNA and detect PCR inhibition using real-time PCR and multiple dye channel fluorescence detection. These include the Applied Biosystems Quantifiler™ Human DNA Quantification, Quantifiler™ DUO and Quantifiler™ Trio kits (Green et al. 2005, Barbisin et al. 2009) and the Promega Plexor® HY (Krenke et al. 2008) and PowerQuant® System kits. The Qiagen Investigator Quantiplex Pro and HYres Kits quantify the total human and male DNA in a sample, detect PCR inhibition, and provide a degradation index (Vraneš et al. 2017, Morrison et al. 2020). While the Quantifiler™ and Plexor® HY kits require approximately two hours, the Quantiplex HYres Kit completes in only an hour. The sensitivity of the commercial kits has improved over time and the newest kit, the Quantiplex HYres kit, is the most sensitive. The Quantiplex HYres Kit is sensitive to <1 pg/µL while the Plexor® HY kit is sensitive to 6.4 pg of total DNA. The Quantifiler™ DUO kit is sensitive to 51 pg of DNA. The PowerSeq® Quant MS System, QuantiFluor® ONE dsDNA System, and QuantiFluor® dsDNA System are recommended for use with the PowerSeq® 45GY NGS kit. The more the DNA is degraded, the fewer loci should be expected to be typed using any DNA typing approach. An accurate determination of the quantity of human DNA in a sample is essential for determining the appropriate input of extracted DNA for NGS. The first step in NGS is library preparation. The extracted DNA can be diluted to achieve the optimal input quantity as instructed by the NGS kit manufacturer (Table 3.1). Low template samples may be used, and input optimized by adding more extract and no water to the library preparation PCR reactions but a lower number of reads and reduced coverage may result if sufficient DNA is not supplied in the reaction.

Table 3.1   Manufacturer's Recommended DNA Input Quantity for NGS

Kit	Recommended Input DNA Quantity (ng)
HID-Ion Ampliseq™ Ancestry Panel	1 ng
HID-Ion Ampliseq™ Identity Panel	1 ng
ForenSeq Signature Prep Kit	1 ng (5 μL of 0.2 ng/μL) genomic DNA, or 2 μL crude lysate (e.g., buccal), or one, 1.2 mm FTA punch
PowerSeq™ 46GY	1 ng (up to 15 μL of 0.2 ng/μL) genomic DNA, or one FTA punch, or one nonFTA punch incubated with PunchSolution™, or 2 μL of extract from swab incubated with SwabSolution™
Precision ID GlobalFiler™ NGS STR Panel v2	1 ng genomic DNA in up to 6 μL (and as little as 0.125 ng)

3.5 Library Preparation

Library preparation is a series of steps to prepare a sample for sequencing using NGS. It begins with the previously extracted, quantified, and diluted sample or a sample from an FTA, or similar, sample collection card for direct amplification. Library preparation has two important features. In the procedure, additional sequences termed adaptor sequences and index sequences are added. The number of steps and required time to perform the library preparation vary greatly by manufacturer and process (Table 3.2). An overview and comparison of the Applied Biosystems HID-Ion Ampliseq™ Ancestry Panel and HID-Ion Ampliseq™ Identity Panel, Verogen ForenSeq™ Signature Prep Kit, Promega PowerSeq™ 46GY, and Applied Biosystems Precision ID GlobalFiler™ NGS STR Panel v2 is shown in Table 3.2. The Verogen and Applied Biosystems library preparation kits are offered in 96 and 384 sample options. Additional NGS kits include Qiagen's QIAseq Investigator Missing Persons SNP panel, QIAseq Investigator ID SNP panel, QIAseq Investigator Global Ancestry SNP panel, and QIAseq Investigator Middle East Ancestry SNP panel.

There are three options for analyzing STR and SNP loci using Applied Biosystems kits. The Applied Biosystems Precision ID GlobalFiler NGS STR Panel v2 contains primer sets targeting thirty-six markers, including the same thirty-one autosomal STR markers, amelogenin sex-determining markers AMELX and AMELY, and three additional sex-determining Y markers (i.e., DYS391, SRY and Yindel).. The Applied Biosystems Precision ID Ancestry Panel targets 165 autosomal markers including fifty-five markers developed by Kenneth Kidd and his group and the SNP*for*ID Consortium and 123 markers developed by Michael Seldin and his team and results in

Table 3.2   Comparison of Steps and Time Required for Library Preparation for Kits from Commercial Suppliers

Library Preparation Kit	Number of Markers	Library Preparation Time (hours)
HID-Ion Ampliseq™ Ancestry Panel	165 (includes aiSNPs)	0.5 hands-on ~18 total
HID-Ion Ampliseq™ Identity Panel	124 (124 iiSNPs)	0.5 hands-on ~18 total
ForenSeq™ Signature Prep Kit	231 (includes 27 aSTRs, 24 Y-STRs, 7 X-STRs, 94 iiSNPs, 22 pSNPs, 56 aiSNPs[a], amelogenin for sex determination)	1.5 hands-on ~ 9 total
PowerSeq™ 46GY	46 (includes 22 aSTR markers, 23 Y-STRs, and amelogenin for sex determination)	None specified by manufacturer
Precision ID GlobalFiler™ NGS STR Panel v2	35 (includes 31 STR markers and 4 sex determination markers)	0.5 hands-on ~18 total

[a] Two ancestry markers also used for phenotype estimation.

average amplicon sizes of less than 130 bp. The Applied Biosystems Precision ID Identity Panel targets 124 autosomal SNPs with a high heterozygosity and low Fixation Index (Fst). The GenPlex™ HID system amplified forty-eight of the fifty-two SNP*for*ID SNPs and amelogenin (Johansen et al. 2013).

The Verogen ForenSeq™ Signature Prep kit offers two options: 152 loci using Primer Set A and 231 loci using Primer Mix B. Primer Set B targets twenty-seven aSTRs, twenty-four Y-STRs, seven X-STRs, ninety-four iiSNPs, twenty-two pSNPs, fifty-six aiSNPs, and amelogenin for sex determination (Jäger et al. 2017). The Promega PowerSeq™ 46GY kit targets twenty-two aSTR markers, twenty-three Y-STRs, and amelogenin for sex determination. A comparison of steps in the library preparation workflows for the Applied Biosystems, Promega and Verogen kits is shown in Figure 3.2. Promega has also introduced the PowerSeq™ Auto/Y System Prototype for integration in a lab's NGS workflow (Montano et al. 2018).

The library preparation steps begin with amplifying and enriching the targets and adding tags, indexes for demultiplexing, and adaptor sequences for flow cell binding and conclude with library purification and normalization. In addition to the samples to be sequenced, a positive control (e.g., 2800 M) and a negative control (no template control) should be processed during the library preparation steps. There are two PCR steps in library preparation with the ForenSeq™ kit. In the first PCR step termed PCR1, a forward tag attached to the forward primer and a reverse tag attached to the reverse primer are added to the target amplicon. These same sequence tags are added to all of the targets. In a second PCR step, PCR2, using the

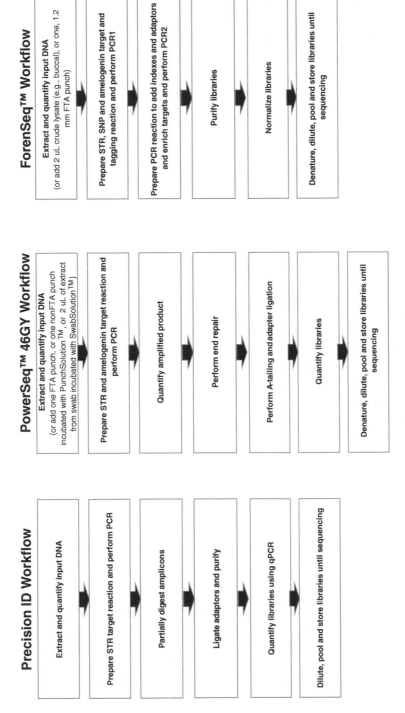

Figure 3.2 Comparison of steps in the library preparation workflows for three NGS products.

tags added in PCR1, adaptor sequences are added adjacent to the primer sequences to allow the amplicons to bind the flow cell for sequencing, and unique forward and reverse i5 and i7 index sequence combinations are added to the targets to label them for demultiplexing interpretation. The adaptor sequences are the same for all samples and are used to complementarily bind the samples to the flow cell oligonucleotide lawn during sequencing. The indexes are a unique set of sequences that are used to assign the data to the sample. An index is analogous to a bar code. Each index is comprised of an eight base pair sequence used to demultiplex the sample data following sequencing. Tables 3.3 and 3.4 list the i5 and i7 index sequences, respectively. Thus the completed "library" for each target consists of an i5 adaptor, i5 index, forward tag, target sequence, reverse tag, i7 index, and i7 adaptor. The ForenSeq PCR1 and PCR2 steps result in amplicons in the 60–460 base pair size range. The amplicon targets and sizes are listed in Chapter 5. The library

Table 3.3   i7 Index Labels and Sequences

Index Label	Sequence
R701	ATCACGAT
R702	CGATGTAT
R703	TTAGGCAT
R704	TGACCAAT
R705	ACAGTGAT
R706	GCCAATAT
R707	CAGATCAT
R708	ACTTGAAT
R709	GATCAGAT
R710	TAGCTTAT
R711	GGCTACAT
R712	CTTGTAAT

Table 3.4   i5 Index Labels and Sequences

Index Label	Sequence
A501	TGAACCTT
A502	TGCTAAGT
A503	TGTTCTCT
A504	TAAGACAC
A505	CTAATCGA
A506	CTAGAACA
A507	TAAGTTCC
A508	TAGACCTA

preparation process has been automated by the French National Police using a Hamilton ID NGS-V STARlet robot (Laurent et al. 2017).

The ThermoFisher Ion Chef™ robot can be used to perform NGS library preparation for any of the human identification (HID) – Ion AmpliSeq panels: GlobalFiler NGS STR Panel v2, Identity Panel, Ampliseq™ Identity Panel, and Ampliseq™ Ancestry Panel as well as the mtDNA Whole Genome Panel and mtDNA Control Region Panels that will be described in Chapter 7. The Ion Chef also performs template preparation and chip loading. Alternatively, the libraries can be prepared manually. If using the robot, the plate with the eight samples to be sequenced, consumables and master mix are all loaded to the Ion Chef™. The consumables include Ion S5 Precision ID Chef Solutions reagent cartridges, chip adapter, enrichment strip cartridge, tip cartridge with pipet tips, PCR plate and frame seal, recovery station disposable lid, recovery tubes, and one or two sequencing chips. A camera system reads the barcodes on the items loaded into the instrument. Three pipetting steps take approximately fifteen minutes and need to be performed manually prior to loading the samples to the platform.

In conclusion, a library is a DNA sample that is ready for sequencing with indexes and adaptors attached to each end. Following manual library preparation, the samples are added to the same tube to form a mixture, or multiplexed, of multiple libraries pooled together that is ready for sequencing.

3.6 Library Purification and Normalization

Following library preparation is the library purification step. Library purification, or clean-up step, removes excess primers and reagents. The ForenSeq™ Signature Prep Kit employs a magnetic bead approach. Working with fewer samples at a time for the bead-based steps leads to more successful sequencing results. In the procedure, the library is first added to a uniform quantity of magnetic beads and the DNA binds the beads. A magnetic block is used to draw the DNA-bound beads to its surface at the bottom of the plate. Ethanol washes are used to wash the beads and remove residual primers and PCR reagents. The excess ethanol is removed, and then the DNA library is released from the beads with resuspension buffer (RSB). Users must be careful to remove all of the ethanol. The library resulting after the purification step results in excess volume of each library than will be needed for sequencing. An agarose or polyacrylamide gel or BioAnalyzer or QIAcel instrument can be used to check the amplicon lengths and quality prior to sequencing. Amplicons in the 60–460 bp range indicate high-quality ForenSeq libraries. Short amplicons of approximately <60 bp indicate primer dimers. Figure 3.3 shows an agarose gel of amplicons prepared using the ForenSeq kit for NGS

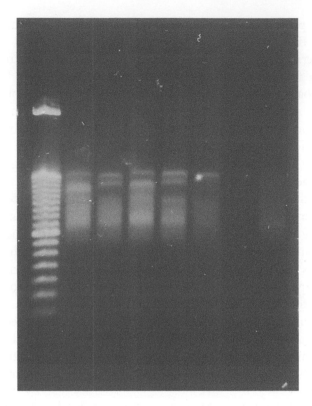

Figure 3.3 Agarose gel of ForenSeq library preparation amplicons (From left to right: Lane 1: Trackit 50 bp ladder with bright bands at 350/800/2500 bp, Lanes 2–6: DNA standards, Lane 7: NTC, Lane 8: DNA standard from 10-month-old library prep). (Courtesy of Adam Klavens.)

library preparation, and Figure 3.4 shows a QIAcel graph of PCR amplicon for pyrosequencing.

As described above, the Ion Chef™ performs not only the library preparation steps but also the library quantitation, purification, library normalization, and chip loading steps.

Following library purification is the library normalization step. The eight-sample library can be quantified using the Ion Library TaqMan® Quantitation kit. Since the PCR steps in library preparation can result in a range of yields, the library normalization step is used to normalize the quantity and concentration of each sample library to ensure that each library is represented equally upon pooling. After it has been determined which samples will be sequenced in the same run, the prepared libraries need to be normalized and pooled together for the sequencing run. Libraries prepared at different times can be normalized together to prepare them for sequencing on the same flow cell as long as they have unique index combinations.

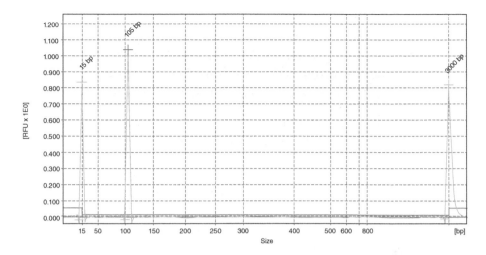

Figure 3.4 QIAcel graph of PCR amplicon for pyrosequencing.

The pooled libraries are diluted to 50 pM and mixed according to the group of barcode adaptors (1-32) for a 530 chip.

The ForenSeq™ Signature Prep Kit employs bead-based normalization. The normalization process uses a standardized quantity of magnetic beads that bind the same amount of library for each sample. The beads must be warmed to room temperature before use for appropriate binding and mixed well to ensure delivery of an even quantity of beads to each library. Beads are added to each well containing library and bind an equal, maximum quantity of each library based upon the binding capacity and quantity of beads added. The magnetic beads bind the DNA libraries, the excess is removed, and the normalized libraries are eluted from the beads.

Verogen indicates that the library preparation step using the ForenSeq™ Signature Prep Kit takes approximately nine hours. In practice, the hands-on time varies with the number of samples and the experience of the investigator preparing the library with the steps and procedures.

3.7 Multiplexing and Denaturation

The last step before sequencing using the ForenSeq™ kit is pooling, or multiplexing, the diluted, normalized libraries. Five microliters of each multiplexed set of targets for each sample to be sequenced are pooled together. A Human Sequencing Control (HSC) is added to the normalized, pooled libraries prior to sequencing, hybridization buffer is added, and the mixture is denatured then snap-cooled on ice. Sequencing the prepared libraries is the focus of Chapter 4.

The Ion Chef™ contains a thermocycler and performs the library preparation, library purification, library normalization, and chip loading steps in an automated, seven-hour process for the eight samples. Primer sequences are removed prior to sequencing using modifications to the sequences. Through the process, each DNA sample is cut into millions of fragments using a mixture of restriction enzymes. The kit uses the Ion Xpress™ Barcode Adapters 1–96 Kit and the IonCode™ Barcode Adapters 1–384 Kit. Barcoded libraries can be combined and loaded onto a single Ion chip to minimize the sequencing run time and cost and allow for accurate sample-to-sample comparisons. Each fragment attaches to its own primer-coated bead, termed templated ion sphere particles (ISPs). The ISPs are purified, and those positive for template are placed at the enrichment station and then loaded onto the chip. Each bead flows across the chip and falls into a well. When using the Ion Chef with a 530 chip, twenty-four samples can be loaded on one chip for autosomal NGS library preparation in aproximately 10 hours.

Questions

1. Are DNA extraction and quantitation required prior to NGS library preparation? Explain why or why not.
2. List some advantages and disadvantages of qPCR DNA quantitation approaches.
3. What components are included in commercial kits for NGS DNA typing?
4. Which controls and standards should be prepared during the sample preparation steps?
5. What is the purpose of the normalization step in library preparation?
6. Describe expected outcomes of sequencing without normalizing the pooled samples.
7. Why are the index sequences added prior to sequencing?
8. What approaches can be used to determine if high-quality libraries were prepared?
9. Which NGS system is best to implement with new users or novices to the process? Explain your answer.
10. Which NGS system reduces interoperator variability? Explain your answer.

References

Barbisin, M., Fang, R., O'Shea, C.E., Calandro, L.M., Furtado, M.R., and J.G. Shewale. "Developmental validation of the Quantifiler Duo DNA Quantification kit for simultaneous quantification of total human and human male DNA and detection of PCR inhibitors in biological samples." *Journal of Forensic Sciences* 54, no. 2 (March 2009): 305–319. doi:10.1111/j.1556-4029.2008.00951.x.

Brevnov, M.G., Pawar, H.S., Mundt, J., Calandro, L.M., Furtado, M.R., and J.G. Shewale. "Developmental validation of the PrepFiler Forensic DNA Extraction Kit for extraction of genomic DNA from biological samples." *Journal of Forensic Sciences* 54 (May 2009): 599–607. doi:10.1111/j.1556-4029.2009.01013.x.

Butler, J. *Forensic DNA Typing*, 2nd ed. Burlington, MA: Elsevier Academic Press, 2005.

Carrasco, P., Inostroza, C., Didier, M., Godoy, M., Holt, C. L., Tabak, J., and A. Loftus. "Optimizing DNA recovery and forensic typing of degraded blood and dental remains using a specialized extraction method, comprehensive qPCR sample characterization, and massively parallel sequencing." *International Journal of Legal Medicine* 134, no. 1 (January 2020): 79–91. doi:10.1007/s00414-019-02124-y.

Castella, V., Dimo-Simonin, N., Brandt-Casadevall, C., and P. Mangin. "Forensic evaluation of the QIAshredder/QIAamp DNA extraction procedure." *Forensic Science International* 156, no. 1 (January 6, 2006) 70–73. doi:10.1016/j.forsciint.2005.11.012.

Desmyter, S., De Cock, G., Moulin, S., and F. Noël. "Organic extraction of bone lysates improves DNA purification with silica beads." *Forensic Science International* 273 (April 2017): 96–101. doi:10.1016/j.forsciint.2017.02.003.

Dukes, M.J., Williams, A.L., Massey, C.M., and P.W. Wojtkiewicz. "Technical note: Bone DNA extraction and purification using silica-coated paramagnetic beads." *American Journal of Physical Anthropology* 148 (July 2012): 473–482. doi:10.1002/ajpa.22057.

Edson, S.M. "Extraction of DNA from skeletonized postcranial remains: A discussion of protocols and testing modalities." *Journal of Forensic Sciences* 64, no. 5 (September 2019): 1312–1323. doi:10.1111/1556-4029.14050.

Elkins, K.M. *Forensic DNA Biology: A Laboratory Manual*. Waltham, MA: Elsevier Academic Press, 2013.

Elkins, K.M., Klavens, A.J. Gorr, K.K., Kollmann, D. D., and C.B. Zeller. "Assessing the performance of next generation sequencing for determining sex, ancestry, and phenotypic characteristics of historic human remains." , in preparation.

England, R., Nancollis, G., Stacey, J., Sarman, A., Min, J., and S. Harbison. "Compatibility of the ForenSeq™ DNA signature prep kit with laser microdissected cells: An exploration of issues that arise with samples containing low cell numbers." *Forensic Science International: Genetics* 47 (July 2020): 102278. doi:10.1016/j.fsigen.2020.102278.

Eychner, A.M., Schott, K.M. and K.M. Elkins. "Assessing DNA recovery from chewing gum." *Medicine, Science and the Law* 57, no. 1 (January 1, 2017): 7–11. doi:10.1177/0025802416676413.

Green, R.L., Roinestad, I.C., Boland, C., and L.K. Hennessy. "Developmental validation of the quantifiler real-time PCR kits for the quantification of human nuclear DNA samples." *Journal of Forensic Sciences* 50, no. 4 (July 2005): 809–825.

Hasap, L., Chotigeat, W., Pradutkanchana, J., Asawutmangkul, W., Kitpipit, T., and P. Thanakiatkrai. "Comparison of two DNA extraction methods: PrepFiler® BTA and modified PCI-silica based for DNA analysis from bone." *Forensic Science International: Genetics Supplement Series* 7, no. 1 (December 2019): 669–670. doi:10.1016/j.fsigss.2019.10.132.

Hoff-Olsen, P., Mevag, B., Staalstrom, E., Hovde, B., Egeland, T., and B. Olaisen. "Extraction of DNA from decomposed human tissue: An evaluation of five extraction methods for short tandem repeat typing." *Forensic Science International* 105, no. 3 (November 8, 1999): 171–183. doi:10.1016/S0379-0738(99)00128-0.

Horsman, K.M., Hickey, J.A., Cotton, R.W., Landers, J.P., and L.O. Maddox. "Development of a human-specific real-time PCR assay for the simultaneous quantitation of total genomic and male DNA." *Journal of Forensic Sciences* 51, no. 4 (July 2006): 758–765. doi:10.1111/j.1556-4029.2006.00183.x.

Jäger, A.C., Alvarez, M.L., Davis, C.P., Guzmán, E., Han, Y., Way, L., Walichiewicz, P., Silva, D., Pham, N., Caves, G., Bruand, J., Schlesinger, F., Pond, S.J.K., Varlaro, J., Stephens, K.M., and C.L. Holt. "Developmental validation of the MiSeq FGx forensic genomics system for targeted next generation sequencing in forensic dna casework and database laboratories." *Forensic Science International: Genetics* 28 (May 2017): 52–70. doi:10.1016/j.fsigen.2017.01.011.

Johansen, P., Andersen, J.D., Børsting, C., and N. Morling. "Evaluation of the iPLEX® Sample ID Plus Panel designed for the Sequenom MassARRAY® system. A SNP typing assay developed for human identification and sample tracking based on the SNPforID panel." *Forensic Science International: Genetics* 7, no. 5 (September 2013): 482–487. doi:10.1016/j.fsigen.2013.04.009.

Kampmann, M.L., Buchard, A., Børsting, C., and N. Morling. "High-throughput sequencing of forensic genetic samples using punches of FTA cards with buccal swabs." *BioTechniques* 61, no. 3 (September 1, 2016): 149–151. doi:10.2144/000114453.

Klavens A, Kollmann DD, Elkins KM, Zeller CB. Comparison of DNA yield and STR profiles from the diaphysis, mid-diaphysis, and metaphysis regions of femur and tibia long bones. J Forensic Sci. 2021 May;66(3):1104–1113. doi: 10.1111/1556-4029.14657. Epub 2020 Dec 28. PMID: 33369740.

Krenke, B.E., Nassif, N., Sprecher, C.J., Knox, C., Schwandt, M., and D.R. Storts. "Developmental validation of a real-time PCR assay for the simultaneous quantification of total human and male DNA." *Forensic Science International: Genetics* 3, no. 1 (December 2008): 14–21. doi:10.1016/j.fsigen.2008.07.004.

Laurent, F.X., Ausset, L., Clot, M., Jullien, S., Chantrel, Y., Hollard, C., and L. Pene. "Automation of library preparation using Illumina ForenSeq kit for routine sequencing of casework samples." *Forensic Science International: Genetics Supplement Series* 6 (December 2017): e415–e417. doi:10.1016/j.fsigss.2017.09.156.

Montano, E.A., Bush, J.M., Garver, A.M., Larijani, M.M., Wiechman, S.M., Baker, C.H., Wilson, M.R., Guerrieri, R.A., Benzinger, E.A., Gehres, D.N., and M.L. Dickens. "Optimization of the Promega PowerSeq™ Auto/Y system for efficient integration within a forensic DNA laboratory." *Forensic Science International: Genetics* 32 (January 2018): 26–32. doi:10.1016/j.fsigen.2017.10.002.

Moore, David, and Dennis Dowhan. "Purification and concentration of DNA from aqueous solutions." Current protocols in pharmacology vol. Appendix 3 (2007): 3C. doi:10.1002/0471141755.pha03cs38.

Morrison, J., McColl, S., Louhelainen, J., Sheppard, K., May, A., Girdland-Flink, L., Watts, G., and N. Dawnay. "Assessing the performance of quantity and quality metrics using the QIAGEN Investigator® Quantiplex® pro RGQ kit." *Science & Justice* 60, no. 4 (July 2020): 388–397. doi:10.1016/j.scijus.2020.03.002.

Naue, J., Sänger, T., and S. Lutz-Bonengel. "Get it off, but keep it: Efficient cleaning of hair shafts with parallel DNA extraction of the surface stain." *Forensic Science International. Genetics* 45 (March 2020): 102210. doi:10.1016/j.fsigen.2019.102210.

Preuner, S., Danzer, M., Pröll, J., Pötschger, U., Lawitschka, A., Gabriel, C., and T. Lion. "High-quality DNA from fingernails for genetic analysis." *The Journal of Molecular Diagnostics* 16, no. 4 (July 2014): 459–466. doi:10.1016/j.jmoldx.2014.02.004.

Sidstedt, M., Rådström, P., and J. Hedman. "PCR inhibition in qPCR, dPCR and MPS-mechanisms and solutions." *Analytical and Bioanalytical Chemistry* 412, no. 9 (2020): 2009–2023. doi:10.1007/s00216-020-02490-2.

Tasker, E., LaRue, B., Beherec, C., Gangitano, D., and S. Hughes-Stamm. "Analysis of DNA from post-blast pipe bomb fragments for identification and determination of ancestry." *Forensic Science International: Genetics* 28 (May 2017): 195–202. doi:10.1016/j.fsigen.2017.02.016.

Vraneš, M., Scherer, M., and K. Elliott. "Development and validation of the Investigator® Quantiplex Pro Kit for qPCR-based examination of the quantity and quality of human DNA in forensic samples." *Forensic Science International: Genetics Supplement Series* 6 (December 2017): e518–e519. doi:10.1016/j.fsigss.2017.09.207.

Walsh, P.S., Mitzger, D.A., and R. Higuchi. "Chelex-100 as a medium for simple extraction of DNA for PCR-based typing from forensic material." *BioTechniques* 10, no. 4 (April 1991): 506–513.

Zeng, X., Elwick, K., Mayes, C., Takahashi, M., King, J.L., Gangitano, D., Budowle, B., and S. Hughes-Stamm. "Assessment of impact of DNA extraction methods on analysis of human remain samples on massively parallel sequencing success." *International Journal of Legal Medicine* 133, no. 1 (January 2019): 51–58. doi:10.1007/s00414-018-1955-9.

Zupanič Pajnič, I., Obal, M., and T. Zupanc. "Identifying victims of the largest Second World War family massacre in Slovenia." *Forensic Science International* 306 (January 2020): 110056. doi:10.1016/j.forsciint.2019.110056.

Performing Next Generation Sequencing

4

4.1 Performing Next Generation Sequencing

The focus of Chapter 3 was on sample preparation steps including DNA sampling and extraction, DNA quantitation, library preparation, clean-up, and normalization. In this chapter, the focus will be sequencing the prepared libraries using the MiSeq FGx or Ion series instruments. The focus on these two instruments results from their adoption for forensic use. They have both proven to be reliable and robust in several years of use in many labs all over the world. Both are very easy to maintain and use.

4.2 Verogen MiSeq FGx® Sequencing

The Verogen MiSeq FGx instrument is a modified Illumina MiSeq platform instrument. The MiSeq is a mid-capacity NGS instrument positioned between the iSeq and MiniSeq benchtop sequencers with less capacity and the HiSeq and NextSeq sequencers with greater capacity and higher throughput. The MiSeq requires the user to supply only the reagent cartridge, a plastic casing with multiple wells each prefilled with the necessary sequencing reagents, flow cell to sequence pooled, normalized libraries, and a waste bottle. The instrument has two run options: research use only (RUO) and forensic genomics (Figure 4.1).

The reagent cartridges and flow cells can only be purchased from Verogen for forensic use while the same from Illumina can be used for RUO mode applications. The flow cell is a modified glass slide outfitted with microfluidic channels. Ninety-six libraries can be sequenced on a standard flow cell using ForenSeq Primer Set A versus thirty-two prepared with Primer Set B. On a microflow cell, thirty-two libraries can be sequenced using ForenSeq Primer Set A, and twelve can be sequenced using Primer Set B.

Unless the sequencer is used frequently, it is best to perform a maintenance wash prior to commencing a sequencing run (Figure 4.2). The maintenance wash consists of three, thirty-minute washes with freshly prepared 0.5% Tween 20. The wells of a plastic cartridge that mimics the shape of the reagent cartridge are filled with the Tween detergent and placed in the instrument compartment. The waste reservoir is emptied, and a used flow cell is

DOI: 10.4324/9781003196464-4   47

Figure 4.1 Setting up a sequencing run on the MiSeq FGx (RUO or Forensic Use).

Figure 4.2 Wash screen on MiSeq FGx.

placed in the flow cell holder in its compartment. The waste bottle and the wash bottle are placed in the main compartment. The wash cycle is begun; after each thirty-minute wash, the user must return to the instrument to replace wash agent and indicate to continue to wash.

Following the wash and upon proper instrument functioning, the reagent cartridge is thawed in a shallow water bath for ninety minutes as indicated by the manufacturer and mixed by inversion prior to use. It can be stored on ice for up to six hours after thawing and prior to a sequencing run. The flow cell is removed from the buffer-containing vial which is shipped in, and wiped dry with care with laboratory wipes to remove all streaks before placing it in its holder in the instrument to click it into place. The instrument records its barcode and checks that the flowcell is unused and not expired. The pooled libraries are denatured, and the human sequencing control (HSC) is added. Subsequently, the libraries are applied to the marked location in the reagent cartridge. The foil covering the well can be pierced with a sterile, DNase-free pipet tip. The manufacturer's recommendation in the ForenSeq kit manual is to add seven microliters of the pooled, normalized libraries (PNL) and HSC mixture to the reagent cartridge for sequencing. However, several labs add more than the recommended volume. Verogen recommends not exceeding thirteen microliters of pooled, normalized libraries and HSC mixture. Air bubbles should be avoided during pooled library addition to the reagent cartridge, but can be tapped out if this occurs. The thawed and mixed reagent cartridge containing the PNL is loaded into its unique cupboard space in the instrument so that its unique Radio Frequency ID (RFID) barcode can be read (it can be manually inputted if not read by the software) (ForenSeq™ DNA Signature Prep Reference Guide).

A new run is created in the Verogen Universal Analysis Software (UAS) prior to beginning the sequencing run (Figure 4.3). It contains the sample names, primer set, and index assignments for each sample. The user also defines the sample type such as sample, positive amplification control, negative amplification control, or reagent blank (NTC). Alternatively, sample information can be imported from a .txt file. Once all samples are added and the attributes have been assigned, the run is saved.

The ForenSeq libraries are amplified and sequenced on the flow cell (Figure 4.4) in the sequencing run. At the instrument computer interface, the

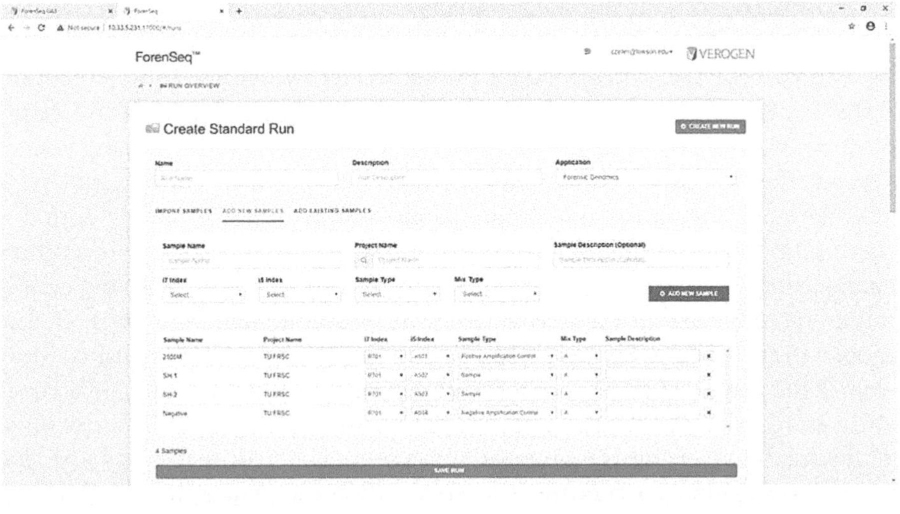

Figure 4.3 Preparing a new run in the Verogen Universal Analysis Software.

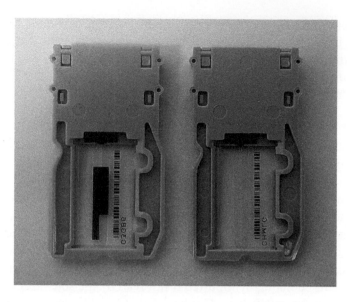

Figure 4.4 Micro (left) and standard (right) MiSeq flow cells.

MiSeq FGx software prompts lead the user through the process of starting the sequencing run. The "sequence" box is followed by "forensic genomics" and then the run name can be selected. The software proceeds with a series of pre-run checks and then the run can be started. Quality information metrics including cluster density, cluster passing filter, phasing, and pre-phasing are reported in real-time as the run progresses. The run status is also visible, and the number of steps are reported (Figure 4.5). The run can be paused or stopped using buttons on this screen.

Within the instrument, the samples are introduced over the flow cell. The flow cell glass slide is covered with lanes populated with an oligonucleotide lawn containing two types of oligos on its surface, which are complementary to the i5 and i7 adaptors added. The sequencing by synthesis (SBS) chemistry proceeds via isothermal amplification. Prior to the actual sequencing occurs a series of steps called cluster generation. A cluster is a distinct spot on a flow cell made of approximately one thousand copies of a library containing one amplicon. The libraries bind to one type of oligo via complementary base pairing via the adaptor region in random locations on the flow cell. A polymerase extends the oligo by adding the base complementary strand. The double-stranded complex is denatured, and the library strand is washed away leaving the extended amplicon of the complementary library sequence covalently attached to the flow cell. The oligo strand folds over to a nearby oligo of the second type and the other adaptor sequence hydrogen bonds so that a U-shaped bridging structure is formed. The polymerase extends the second oligo by complementary base pairing to the oligo strand through bridge

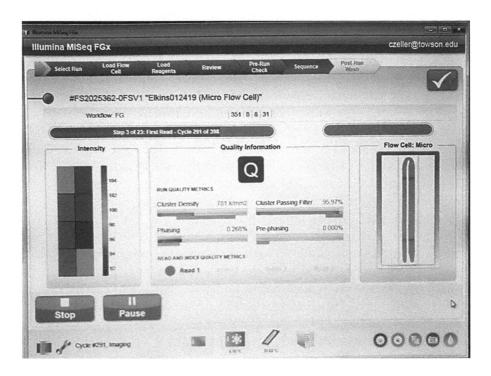

Figure 4.5 Sequencing in process on the MiSeq FGx.

amplification. The bridge is denatured and two single-stranded copies of the library are now covalently attached and extended from the oligo lawn. The process is repeated simultaneously in parallel so that billions of clusters are generated. The reverse strands are cleaved and washed away leaving only the forward strands. The 3′-ends are blocked to prevent unwanted priming.

Finally, the sequencing begins with Read 1. A sequencing primer complementary to forward tag, which was added in PCR1, initiates sequencing on the forward strand. One of four fluorescently labeled nucleotides is added to the primer based upon the complementary sequence of the strand in a massively parallel sequencing (MPS) process. An ultraviolet light source is used to excite the fluorophore and enable the detection of each cluster. After the addition of each base, the emission wavelength is used to determine the nucleotide base that was added. The fluorescent dye is cleaved off, and the blocking agent is removed so that the next base can be added. The process continues for the full length of the strand. The fluorescence emission of the clusters over the flow cell is read simultaneously generating hundreds of high-quality images that require significant storage space. After the forward strand first read sequence is completed, the product is washed away. Next in Read 2, the index 1 i7 read primer is hybridized to the forward strand, and sequencing is conducted again until the strand is completed.

When sequencing is completed, it is washed away. Then, the 3′-end of the template is deprotected, folds over, and binds the second oligo on the flow cell. Extension of the second oligo yields a read of index 2 i5 in Read 3. The index 2 product is washed off after the sequencing and detection are complete. Finally, the second oligo is sequenced fully until the double-stranded bridge is formed. The bridge is denatured and linearized, and the forward strand is cleaved off and washed away. A read of the target reverse strand is begun with the addition of a sequencing primer. The read continues until the desired length of the second strand is sequenced (Launen 2017). Then the product is washed away. The sequencing run takes twenty-seven to thirty hours. The run completion details are displayed on the screen and can be viewed in the UAS (Figure 4.6).

Illumina recommends keeping the MiSeq powered on at all times with frequent use unless the shutdown steps are followed to power down the instrument for long periods of disuse. At the conclusion of the sequencing run, the used flow cell, wash tray, wash bottle, and waste bottle can remain in place until the next run. An instrument wash should be performed, however, immediately before and after each run. There are three wash options: a post-run wash, a maintenance wash, and a standby wash. A wizard leads the user through the wash steps. A post-run wash takes thirty minutes and 25.5 mL of wash solution. The wash solution contains DNase-free and RNase-free water mixed with Tween 20 resulting in a 0.5% Tween 20 final concentration. Dilute bleach solution is used in position seventeen of the wash tray, corresponding to the position of library addition. The maintenance wash consists of a series of three wash steps as described above. It should be performed at least once

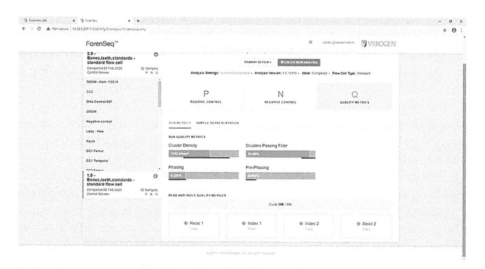

Figure 4.6 MiSeq FGx sequencing run completion viewed in UAS.

every thirty days. A maintenance wash uses 17.5 mL of wash solution and takes approximately ninety minutes. According to the manufacturer, the standby wash should be used if the instrument will not be used within seven days and should be repeated every thirty days that the instrument is idle. The standby wash uses 46 mL of wash solution and takes approximately two hours. Prior to shutting the instrument down, a maintenance wash should be performed, the waste should be discarded, and the waste bottle should be returned to its compartment. To shut down the instrument, select "shut down" from the Manage Instrument screen and toggle the power switch to off. The user should wait at least sixty seconds before restarting the instrument. Toggle the power switch on and allow the chiller compartment to cool before using the instrument (ForenSeq™ DNA Signature Prep Reference Guide).

4.3 ThermoFisher Ion Torrent™ and Ion PGM Sequencing

The ThermoFisher Ion GeneStudio S5 System, Ion PGM™ System, Ion S5™ XL System, and Ion OneTouch™ 2 System sequencing systems can be used to sequence libraries prepared using the Applied Biosystems Precision ID GlobalFiler™ NGS STR Panel v2, HID-Ion Ampliseq™ Identity Panel, or HID-Ion Ampliseq™ Ancestry Panel kits with the Ion 510, Ion 520, and Ion 530 semiconductor chips. The higher chip numbers have a higher capacity for sequencing libraries to be sequenced on the chip. The Ion 520 chip can accommodate sixteen sample libraries prepared with the GlobalFiler kit. The Ion 530 chip can run thirty-two sample libraries prepared with GlobalFiler™ kit. For the Ancestry Panel, 48 samples can be run on the Ion 510 chip, 72 can be run on the Ion 520 chip, and 362 samples can be run with the Ion 530 chip. For the Identity Panel, 54 samples can be run on the Ion 510 chip, 81 can be run on the Ion 520 chip, and 384 samples can be run with the Ion 530 chip. The sample is loaded directly into the chip well.

To begin sequencing with the Ion Torrent system, the chip, reagent cartridge, wash solution, sequencing buffer, and waste bottle must be loaded to the instrument in the initialization proces. The reagent cartridge must be equilibrated at room temperature prior to use and the wash solution must be inverted several times. The chip, cleaning solution, sequencing buffer, and waste container are loaded into one compartment. The chip is attached to a bar and slides in its drawer. Each of these items is tracked using an RFID system. The reagent cartridge contains all of the nucleotides that can be added during sequencing. The instrument initializes when these items have been loaded. Before beginning the sequencing run, the user designs a sequencing protocol that specifies the number of barcodes, chip type, chip barcode, run module (e.g., Global Filer/Ancestry/whole mtDNA) and the number of flows.

The sequencing protocol parameters can vary with the sequencing to be performed by the instrument

The chip prepared using the Ion Chef is placed into the Ion S5 or Ion S5 XL. The chip has millions of wells covering pixels for detection. The user selects the "home" then "run" option to begin. The sample is loaded directly into the chip well, and the new chip loaded with ISPs is placed in the sequencer for sequencing and engaged with the chip clamp. Once the protocol is created as described above, the sequencer automatically detects the chip barcode and performs sequencing (Precision ID GlobalFiler™ NGS STR Panel v2 with the HID Ion S5™/HID Ion GeneStudio™ S5 System Application Guide).In a series of steps, the chip is flooded with one of the four natural DNA nucleotides. The Watson-Crick base pair is formed by hydrogen bonding when the appropriate base binds the template. Upon incorporation of the base into the chain, a proton is released changing the pH of the system. The pH change is detected by an ion-sensitive layer in the chip, and the voltage is detected by the detector. The voltage is converted into a digital signal, and the base call is made by the software. The chip is flooded with a new base every fifteen seconds. If the base is not complementary to the next base, no hydrogen ion is released, no voltage change is detected, and no base call is made. If two (or three) identical bases are next to each other, the voltage will increase by double (or triple) based upon the number of nucleotides incorporated. Each base is called for the sequences extended on the millions of beads in the millions of wells. Each sequencing run requires three and a half hours. Following the run, users should perform a post-run clean.

4.4 The Next Step

Following DNA sequencing, the run and sequence files are evaluated to determine the quality of the run. The data is analyzed using a variety of tools. These steps are the topic of the next chapter.

Questions

1. Explain the concept of sequencing by synthesis.
2. Explain how an added base is detected on the MiSeq FGx and Ion series instruments.
3. Why are the consumables labeled with barcodes?
4. List some reasons why a sequencing run may fail.
5. What information must be inputted into the sequencer software for interpretation after sequencing?

References

ForenSeq™ DNA Signature Prep Reference Guide. August 2020. Accessed May 21, 2021. https://verogen.com/wp-content/uploads/2020/08/forenseq-dna-signature-prep-reference-guide-VD2018005-c.pdf.

Launen, L. "Illumina Sequencing (for Dummies) – An overview on how our samples are sequenced." Accessed January 22, 2021. https://kscbioinformatics.wordpress.com/2017/02/13/illumina-sequencing-for-dummies-samples-are-sequenced/.

Precision ID GlobalFiler™ NGS STR Panel v2 with the HID Ion S5™/HID Ion GeneStudio™ S5 System Application Guide. Revision 15 November 2018. Accessed May 21, 2021. https://assets.thermofisher.com/TFS-Assets/LSG/manuals/MAN0016129_PrecisionIDSTRIonS5_UG.pdf.

Next Generation Sequencing Data Analysis and Interpretation

5

5.1 NGS Data Analysis

Next generation sequencing (NGS) produces large quantities of data – exponentially more than traditional methods for STR typing and SNP analysis. The data output is also different. Whereas fluorescence intensity data is recorded by CE instruments, the analogous NGS output is the number of sequencing reads. Some NGS instruments record photographic images of the raw data, which are in fact fluorescent dots of clusters on the chip. Software is used to interpret the fluorescence emission or voltage changes and generate a raw DNA sequence for each cluster. The number of sequences of each type is counted, and tables are generated showing the number of "reads" or counts of each sequence recorded. As previously discussed in Chapters 2 and 4, NGS instruments use several different technologies to record the identity of the nucleotide base as it is incorporated. While determining the STR repeat number and SNP variant base is an aspect of both CE and NGS data interpretation, there are several new terms that are used to qualify NGS data that are not applicable to CE. Some are platform-specific and others are more general to scoring large sequence data sets.

The term cluster density (K/mm^2) refers to the number of individual "islands" or groups of DNA molecules that were amplified into clusters on a flow cell. Each cluster represents a single, unique template on the Illumina platform. The depth of coverage indicates the average number of times a sequence is recorded in the process. In NGS protocols, the genome is fragmented or the targets are preferentially amplified and tagged prior to sequencing, thus the depth of coverage is an indicator of the strength of the data obtained. The clusters passing filter (%) indicates percentage of individual clusters the software is able to distinguish. The filter removes the least reliable data, such as that from overlapping clusters. Phasing is a PCR-based phenomenon that occurs when a base fails to be added in a sequencing cycle. Phasing is detected upon comparing the subsequent base sequence addition to other clusters. Prephasing is another PCR-based issue and occurs when an extra base is added in a cycle leading to a read one base longer than the other libraries. This is also detected upon sequence comparison.

DOI: 10.4324/9781003196464-5

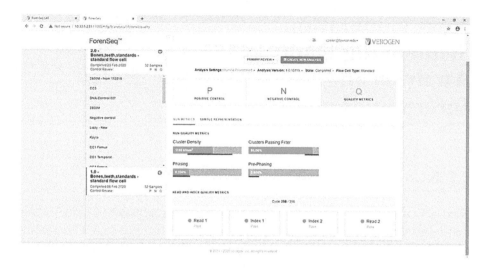

Figure 5.1 MiSeq FGx run metrics for a successful sequencing run.

Typically only a small portion of the strands in a cluster, if any, become out of phase. A screenshot of the Illumina MiSeq FGx run metrics viewed in the UAS is shown in Figure 5.1.

The Fastq file is the text-based output of the nucleotide-based sequence and can be analyzed by many software applications. The Q-Score is a quality measure of $Q = -10\log 10_e$ where e is the estimated probability that the base call is wrong. The Q-score is an estimate of the quality of the base call using a log function. The quality score is based on the Phred score from Sanger sequencing. The Phred score is a rating of the quality of the nucleotide base identification in Sanger sequencing. High Q scores indicate a low probability that the base call is incorrect, while a low Q score reflects low quality or unusable data.

5.2 Verogen Universal Analysis Software

The Illumina MiSeq and the Verogen MiSeq FGx record a photographic image of all of the fluorescing clusters on the flow cell after the addition of the base in each cycle. A camera is used to record the fluorescence signal after each base is added at each cluster location to produce the base read. (A cluster is needed so that the signal is strong enough to be detected by the detector.) A good run will reflect clusters that are evenly spaced across the flow cell. Poor flow cells will exhibit blank areas where clusters were not generated or images showing many overlapping clusters. These images are large and collectively represent the majority of megabytes of data that is recorded in each sequencing run.

The MiSeq FGx outputs metrics for cluster density, percent of clusters passing the filter, and phasing and prephasing that indicate the quality of the run and issue warning flags when the values are outside the set range. These metrics and the percentage of the reads completed are shown in the main run window. The metrics are color-coded green if the run values are within normal tolerance, orange if a metric is outside of the manufacturer's target range, and red if the indicators reflect major problems with the run. The setting for the clusters passing filters is ≥80%. A green light reflects the values at or above that tolerance. The phasing filter is set at ≤0.25%, and the prephasing filter is set to ≤0.15%. The main window also shows the flow cell coverage and has a stop and pause icon and a quality indicator icon. A passing quality is indicated by a green Q icon and green circles for the read 1, index 1, index 2, and read 2 indicators (Figure 5.1). When the quality is below the threshold, the circle turns orange. If all of the metrics are green, the run is of high quality to proceed with data analysis. Approaches for troubleshooting if signal warnings are indicated are covered in Chapter 6.

Data analysis can be performed with one of several software applications. The software may be loaded to the local server or be accessed via the cloud. For research applications, Illumina offers the Sequencing Analysis Viewer on the cloud to view and analyze the sequencing data. Additional analyses are traditionally conducted using a variety of applications and scripts on BaseSpace. For example, the Q-score can be computed. A Q-score of Q30 indicates a high-quality read or that the base call is 99.9% accurate (a 1:1000 chance that the base is incorrect). Additional sequence analysis tools are described in Section 5.6.

Results from sequencing libraries prepared using the ForenSeq kit (Table 5.1) can be analyzed using the Verogen ForenSeq Universal Analysis Software (UAS) or outside of the UAS using BaseSpace or one of many third-party apps. After the run, the sequencing data is automatically saved on the instrument, copied to the server, and imported into the UAS. The UAS enables initial "secondary analysis" and optional analyses and reporting. Forward and reverse reads are paired to create contiguous sequences which are aligned to the reference genome. The paired-end reads are used to resolve ambiguous alignments. Multiple samples are sequenced together; demultiplexing is a bioinformatics-based approach to identify the index sequences and label them with the appropriate sample name. Thereafter, each sample can be analyzed separately. The UAS demultiplexes the data using the user-defined indexes for each sample, generates the raw sequence files, makes base calls at each locus, and assigns the read to the appropriate STR or SNP based upon counting from the end of the primers. For example, using

Table 5.1   ForenSeq and Precision ID Target Autosomal and Sex Chromosomal STR Loci

Marker	Repeat	Chromosome	Database	Precision ID Amplicon Size (bp)	ForenSeq Amplicon Size (bp)
D1S1656	TAGA	1	Expanded CODIS	163–211	133–192
TPOX	AATG	2	CODIS	167–199	61–109
D2S441	TCTA	2	Expanded CODIS	163–195	137–177
D2S1338	TGCC/TTCC	2	Expanded CODIS	126–190	110–203
D3S1358	TCTA/TCTG	3	CODIS	129–177	138–194
D4S2408	ATCT	4	Non-CODIS	167–191	98–118
FGA	CTTT/TTCC	4	CODIS	137–299	150–312
D5S818	AGAT	5	CODIS	137–169	98–162
CSF1PO	AGAT	5	CODIS	143–183	72–120
D6S1043	AGAT/AGAC	6	Non-CODIS	163–227	154–226
D7S820	GATA	7	CODIS	150–186	118–183
D8S1179	TCTA/TCTG	8	CODIS	151–199	82–138
D9S1122	TAGA	9	Other Autosomal STRs	NA	104–132
D10S1248	GGAA	10	Expanded CODIS	155–199	124–176
TH01	TCAT	11	CODIS	129–173	96–140
vWA	TCTA/TCTG	12	CODIS	147–207	135–195
D12S391	AGAT/AGAC	12	Expanded CODIS	149–193	229–289
D13S317	TATC	13	CODIS	149–181	138–186
PentaE	AAAGA	15	Other STR	168–273	362–481
D16S539	GATA	16	CODIS	135–175	132–184
D17S1301	AGAT	17	Other Autosomal STRs	NA	130–154
D18S51	AAGA	18	CODIS	156–232	136–272
D19S433	AAGG/TAGG	19	Expanded CODIS	155–195	148–240
D20S482	AGAT	20	Other Autosomal STRs	NA	125–157
D21S11	TCTA/TCTG	21	CODIS	179–245	147–265
PentaD	TCTTT	21	Other STR	139–204	209–298
D22S1045	ATT	22	Expanded CODIS	168–201	201–245
SRY	N/A	Y	Sex determination	119	NA
DYS391	TCTA	Y	Sex determination	130–162	119–163
AMEL-X	indel	X	Sex determination	102	NA
AMEL-Y	indel	Y	Sex determination	108	NA
rs2032678	indel	Y	Sex determination	178–183	NA
DYF387S1	[AAAG]n GTAG [GAAG]n [AAAG]n GAAG [AAAG] n [GAAG]n [AAAG]n	Y	Sex determination	NA	207–263

(Continued)

Table 5.1 (*Continued*) ForenSeq and Precision ID Target Autosomal and Sex Chromosomal STR Loci

Marker	Repeat	Chromosome	Database	Precision ID Amplicon Size (bp)	ForenSeq Amplicon Size (bp)
DYS19	TAGA	Y	Sex determination	NA	269–309
DYS385a-b	GAAA	Y	Sex determination	NA	232–316
DYS389I	[TCTG] [TCTA] [TCTG] [TCTA]	Y	Sex determination	NA	236–268
DYS389II	[TCTG] [TCTA] [TCTG] [TCTA]	Y	Sex determination	NA	283–323
DYS390	[TCTA] [TCTG]	Y	Sex determination	NA	290–334
DYS392	TAT	Y	Sex determination	NA	318–362
DYS437	[TCTA]n [TCTG]n [TCTA]n	Y	Sex determination	NA	194–226
DYS438	TTTTC	Y	Sex determination	NA	129–179
DYS439	AGAT	Y	Sex determination	NA	167–211
DYS448	AGAGAT	Y	Sex determination	NA	330–402
DYS460	ATAG	Y	Sex determination	NA	348–376
DYS481	CTT	Y	Sex determination	NA	129–174
DYS505	TCCT	Y	Sex determination	NA	162–186
DYS522	CTTT	Y	Sex determination	NA	298–334
DYS533	ATCT	Y	Sex determination	NA	186–226
DYS549	GATA	Y	Sex determination	NA	210–226
DYS570	TTTC	Y	Sex determination	NA	142–206
DYS576	AAAG	Y	Sex determination	NA	163–223
DYS612	[CCT]5[CTT] [TCT]4[CCT] [TCT]25	Y	Sex determination	NA	275–296
DYS635	TSTA compound	Y	Sex determination	NA	242–302
Y-GATA-H4	TAGA	Y	Sex determination	NA	159–187
DXS10074	AAGA	X	Sex determination	NA	184–244
DXS10103	[TAGA]n [CTGA]n [CAGA]n [TAGA]n [CAGA]n [TAGA]n	X	Sex determination	NA	157–185
DXS10135	[AAGA]n GAAA gga [AAGA]n [AAAG]n	X	Sex determination	NA	239–312

(Continued)

Table 5.1 (*Continued*) ForenSeq and Precision ID Target Autosomal and Sex Chromosomal STR Loci

Marker	Repeat	Chromosome	Database	Precision ID Amplicon Size (bp)	ForenSeq Amplicon Size (bp)
DXS7132	TAGA	X	Sex determination	NA	175–211
DXS7423	[TGGA]n aggacaga [TGGA]n	X	Sex determination	NA	188–220
DXS8378	ATAG	X	Sex determination	NA	434–458
HPRTB	ATCT	X	Sex determination	NA	193–229

n represents number of repeats. NA is not applicable.
Repeats recorded from kit manuals and strbase.nist.gov.

the STR or SNP aligner, the data reads are aligned, converted to allele length (for STRs), and summed for SNP calls based upon comparison to a reference sequence.

Errors with index assignments, sample name or user-inputted information can be corrected, and the data can be reanalyzed. The original data, Version 1.0, is retained when the new analysis is created in the Project with a New Run Version identified as 2.0. All versions of the data are retained as separate files. The analytical threshold (AT) can be retained from the manufacturer (Jäger et al. 2017) or set to −2 bp stutter as well as N+1, N−1, and N−2. Stutter artifacts are PCR misincorporation and sequencing errors. The interpretation threshold (IT) can also be adjusted. The data can be reanalyzed with new threshold settings such as stutter filter, intralocus balance, analytical threshold, and interpretation threshold. Minor setting changes such as these are also saved in a new file (e.g., Version 1.1).

Reads/noise is impacted by low template/input quality and proportional to sample input quantity. Different loci can have different noise levels. In choosing a threshold, the scientist must consider Global AT, Kit/Locus-dependent AT and evaluate which to use by impact on the likelihood ratio (LR). The sample history is recorded for each action taken by the user, and user actions or system events can be viewed at the analysis, sample, and locus level. The sample history can be toggled on and off using a switch. System events include run completion indicators, analysis completion, and completion of population statistics. User events include genotype edits, comments, and sample reports generated by the user.

The UAS displays the data in several graphs and charts. Sequencing differences are identified, and the results are displayed pictorially for each locus of interest. The P representing the positive amplification human sequencing control (HSC) is displayed with a pass or fail metric based upon the overall intensity (Figure 5.2). The N representing the negative control will display the number of STRs and SNPs typed when selected (Figure 5.3). For each

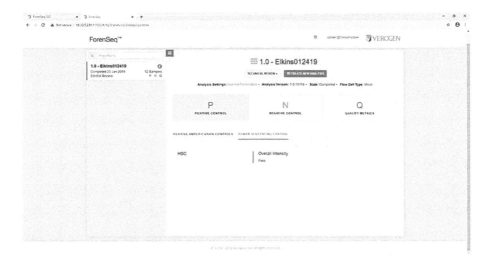

Figure 5.2 Passing HSC in MiSeq FGx sequencing run.

Figure 5.3 ForenSeq sequencing run negative control with no alleles.

sample, a graph is displayed that shows the number (intensity) of reads versus the length of each locus in base pairs (Figure 5.4). A high-quality result for the positive controls and samples will have a high number of reads for all loci, regardless of length (Figure 5.5). Degraded or low template samples often exhibit fewer reads for longer amplicons, and the shorter amplicons are more likely to lead to success in sequencing (Figure 5.6). Table 5.1 shows the amplicon lengths at the loci targeted by the ForenSeq and Precision ID kits. The total number of STR and SNP loci typed for each sample is displayed under the graph in Figure 5.6.

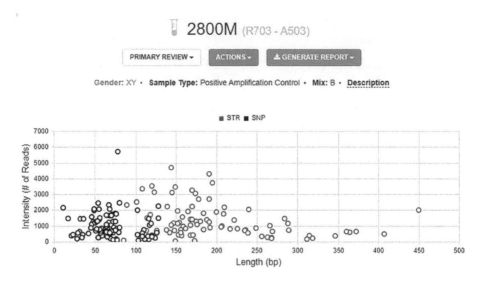

Figure 5.4 UAS reads versus length graph for 2800M.

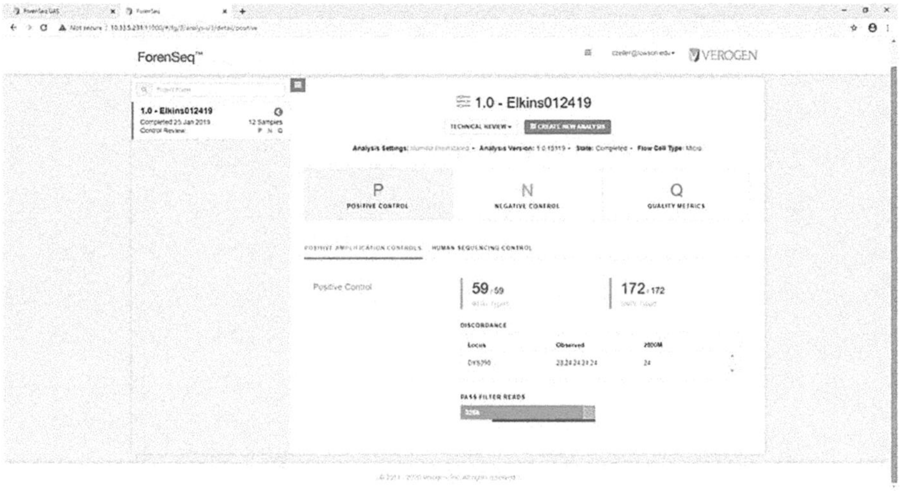

Figure 5.5 Full profile for 2800M using ForenSeq library prep.

The UAS employs several quality indicators to alert the user to issues with the data quality. The potential data quality issues include stutter, extra or missing alleles, read imbalance, low coverage, and alleles below the thresholds. User edits and alleles not detected also cause loci to be flagged. The stutter flag indicates that an "allele" is likely a stutter amplicon of another allele at the locus (Figure 5.7). The allele count flag turns on if more than two alleles have counts above the IT indicating a possible mixture. The imbalance flag indicates

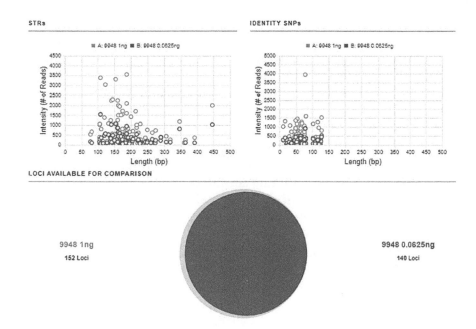

Figure 5.6 Sample comparison in UAS for 9948 at two input concentrations.

Figure 5.7 D7S820 locus displaying stutter and peak imbalance for a sample.

that the read count ratio is below the intralocus balance setting applied by the manufacturer or the user. The low coverage flag indicates that the number of reads is below the IT. The interpretation threshold flag alerts the user when at least one allele is above the analytical threshold but below the interpretation threshold. The analytical threshold flag turns on when the number of reads at the locus is below the analytical threshold, and no alleles were above the interpretation threshold. If no signal is detected for a locus, the not detected flag is activated. The user modified flag is activated if the user manually edits an allele.

A screen displays all of the STR loci in boxes with the numerical allele calls (grey boxes) for each locus and quality indicators (orange boxes), if any. Selecting each box will display the allele call, a slide switch that turns the allele call on or off, the number of reads or intensity for each allele, and detected stutter and the actual sequence of the allele. Being able to evaluate the full STR sequence is a significant advantage of NGS over CE STR typing. A graph displays the number of reads versus alleles with thresholds and color coding to indicate if the alleles are above (blue) or below (grey) the analytical (dark grey) and interpretation thresholds (light grey). An allele that is above the analytical threshold but below the interpretation threshold is displayed in brown. All of the SNP loci are also displayed in boxes showing the heterozygote or homozygote allele calls at each locus (Figure 5.8). SNP imbalance is displayed with a larger font and can be viewed for each locus in a pie chart (Figure 5.9). Selecting a box shows switches to turn an allele call on or off, the total number of reads for each allele and percent of total reads, and a pie graph that displays the alleles.

If more than five loci have more than two STR alleles or imbalance, the results are flagged for mixture analysis. Similarly, if more than ten STR loci are detected to be imbalanced, the data is flagged and a mixture may be detected. Poor data is also flagged for both STR and SNP data. If poor data is detected for samples, the controls can indicate if the run or samples were of

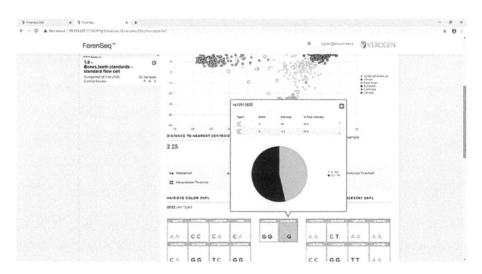

Figure 5.8 Relatively Balanced rs12913832 SNP reads for a sample.

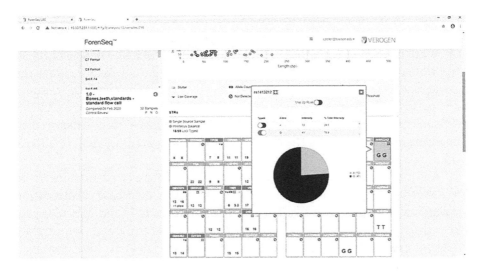

Figure 5.9 Imbalanced rs1413212 SNP reads per sample.

poor quality. For example, a run in which the positive control standard ran successfully and a full profile is produced, but one or more of the samples produce(d) a partial profile or flagged loci would indicate that the issues are isolated to the sample(s). If the standard and samples fail to produce a full profile, this indicates that the issues are global to the run and not necessarily with the samples. The negative controls should have a low or zero read count, and ideally, the samples should have a high read count.

An additional analysis that can be performed in the UAS is population statistics. Computed population statistics indicate the probability that the sample will match an individual at random in a given population. It is not the probability that the sample will match an individual in the given population as the sample may not match any individual in the population. Two methods of computing population statistics are available: Random Match Probability (RMP) and LR. The typed samples can also be directly compared with the results displayed in a Venn diagram. Only loci that are typed in both samples are included in the comparison. Typed STRs and typed iSNPs are compared and population statistics can be added for the compared loci. Upon selecting the box for each locus, the allele call, thresholds, number of reads, and allele sequences are shown side-by-side for both samples. The software will generate an analysis and display discordant STRs and SNPs and show the number of intersecting loci. Discordant STR and SNP loci will be shown in boxes below the number. For the population statistics computations, the loci used in the calculations must have the correct number of alleles (e.g., one or two for autosomal loci), and the sex chromosome loci must have the appropriate number of alleles based upon the called sex using the amelogenin gene.

The UAS can be used to create Project Level Reports including an Autosomal STR Genotype Report with tabulated allele calls at each locus. These are important because CODIS accepts only the allele repeat number. Additionally, a Sample Genotype Report can be ordered which includes the actual repeat sequence for each typed or all alleles at a locus with the number of reads for each. This report includes data for the autosomal STRs, Y STRs, X STRs, iSNPs, sample history, and settings. The Sample Genotype Report also includes graphs for each of the categories above, autosomal STRs, Y STRs, X STRs, and iSNPs.

Additional or tertiary analysis phenotype and ancestry estimation can be performed for samples to aid in missing persons or cold case identification or generate investigative leads. The UAS can perform hair and eye color predictions based upon HIrisPlex loci data from the literature and can estimate biogeographical ancestry using another SNP panel (Liu et al. 2009, Sampson et al. 2011, Nievergelt et al. 2013, Walsh et al. 2011, 2014). For the UAS to perform the hair and eye color prediction, all of the HIrisPlex loci must be typed. The phenotype estimation includes a probability for brown, red, blond, and black hair color and intermediate, brown, and blue eye color. Figure 5.10 shows the prediction for the standard 2800M. The biogeographical ancestry tool was trained with 1000 genome project data (The 1000 Genomes Project Consortium 2012, 2015). In addition to the phenotype prediction, the biogeographical ancestry (BGA) can be predicted. Two component principal component analysis is used to estimate the sample BGA origin. Samples are estimated into one of the following categories: Ad Mixed American, African, East Asian, Eurasian, or may locate to a region between or outside of these

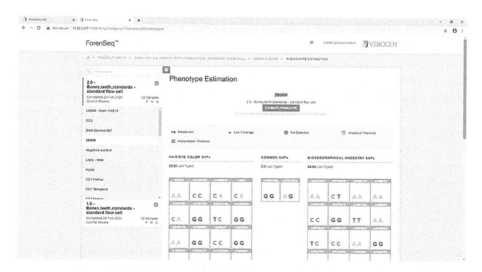

Figure 5.10 UAS phenotype estimate for 2800M.

regions on the graph. The distance to the nearest centroid is also given. There is no minimum number of SNPs required for this prediction. More detail is provided in section 5.4.

Finally, the STR and flanking regions can also be analyzed to discover new, unreported variations and/or additional sources of difference between two very similar samples that were previously undetected by CE as the length remains the same. The flanking regions are the genetic sequence between the PCR primers and the STR repeat region or SNPs of interest. The flanking region can be used for differentiation and deconvolution. These alleles of the same length exhibiting variation in the sequence are termed isoalleles. The MICM method for calculating match probabilities using forensic NGS data trims the data to align to the data in the database. ForenSeq sequences include 15 bp on the 3′ end. The user may need to sequence the flanking regions using Sanger sequencing for the database so that the complete NGS data set can be used in database and frequency calculations. Generating the flanking region report results in an Excel file output. To execute this feature, the project must be analyzed with "flanking regions" enabled in the analysis method. The SNP flanking region report includes the "rs" identifier for the locus congruent with SNPedia and reference of variant SNPs.

To ameliorate some limitations with the UAS, the French National Police developed Programme d'Interprétation Résultats d'Analyses NGS Hautement Amélioré (PIRANHA) in R using the UAS summary report (xls) text file to further interpret ForenSeq data.

5.3 ThermoFisher Converge Software

After an Ion series sequencing run is completed, the data analysis begins immediately at the local Ion Reporter server. Analysis of sequencing data from libraries prepared using the Precision ID GlobalFiler NGS STR v2 Panel yields the number of STR allele repeats and the base sequence for each repeat. Analysis of the sequencing data from libraries prepared using the HID-Ion Ampliseq™ Ancestry Panel and HID-Ion Ampliseq™ Identity Panel yield the SNP base calls at the loci. The Torrent Server records and stores the sequencing data. The data remains on the server, and the data can be viewed on the Ingenuity Variant Analysis in ThermoFisher Ion Reporter™ Software and server. Alternatively, the data can be stored on the cloud and analyzed remotely by pointing the browser to 10.65.1.14. The front page of the Torrent Browser has the login screen and four tiles: plan, monitor, review, and export. Collaborators can access the data via the browser. The local Ion Reporter Server System can be used if the lab does not want data in the cloud to comply with the Health Insurance Portability and Accountability Act of 1996 (HIPPA). The software demultiplexes and sorts the data.

A run report is generated for each run. The report includes metrics such as Ion Sphere™ Particle (ISP) Density, ISP Summary, total bases, total reads, usable reads, key signal, and, on the bottom, an alignment to a human genome sequence, hg19. The ISP Density is a heat map that shows the relative amounts of DNA that were loaded on to the semiconductor chip for sequencing. Red indicates that an excess of DNA was loaded into the wells on the chip. Some DNA loading is represented by yellow while green and blue indicate almost no loading. The loaded area of the PGM chip will be in the shape of a triangular prism. The corners will always be blue indicating no ISPs. The ISP Summary bars indicate how many of the wells were loaded with ISPs. The third bar is clonal and polyclonal fragments; a typical value is 70%. The final bar is the usable sequence library. A histogram of reads and lengths can be exported as a pdf.

Next, once the metrics have been evaluated, the SNP calls can be viewed using the links to the HID_SNP_Genotyper plugin. Summary information about the run is listed at the top. The sample name and barcode ID are shown on the left and results are displayed on the right. Users can download the targets and hotspots as BED files and the allele coverage data as a csv or pdf file. The results section has tabs to display population stats and compare profiles. Within the population stats tab are three tabs: map, results, and genotypes. In the Admixture prediction box, the map view shows the highlighted Admixture prediction map ancestry of the person in question based upon the data collected using the 151 aiSNPs. The results tab shows the population and percentage and a bell curve of the log likelihood with the confidence level of the prediction. The third tab, Genotypes, displays a table that lists the SNPs and genotype. The population likelihoods dropdown box shows the hotspot likelihood on a map, and the results list the population, geographical region, and RMP for the population groups stored in the database based upon the allele frequencies programmed into the plugin. The plugin can be customized with allele frequencies from another data set as well. As with any tool of this kind, the prediction is only as accurate as the database it uses. Although the ethnic groups for each region are broken out, the cluster of groups from a region lends confidence in the region prediction. Scrolling further down in the plugin is allele coverage with table and chart tabs and how many SNPs failed QC checks. The table includes the chromosome number, position, rs number, number of reads coverage, number of reads for each base, genotype, and number of reads on the forward (positive) and reverse (negative) strands. The user can customize the plugin settings to set how many reads are required to make a call and read imbalances. The chart tab displays the data in graphical form. The user can scroll to the region of interest and show the SNPs in that region. The stacked bar graph shows the allele calls at each locus, each with a unique color, and the vertical axis is the coverage. Hovering the mouse over each bar shows the rs number, total reads, genotype, and number of reads for each base and percent. The graph can also be altered to display the results by coverage. The chart and data table can be downloaded to Excel or another software program.

The HID_STR_Genotyper plugin includes a profile summary, coverage plots, and locus data and enables users to evaluate the STR typing results. The profile summary is a clickable list of every sample in the run and an STR allele table similar to a GeneMapper report table. Highlighting the row populates the charts below. The default view is analogous to CE run data with peaks of varying heights for the alleles. The STR allele subtypes are displayed with vertical bar graphs with the number of reads labeled on each. The peak height reflecting the read coverage is analogous to the relative fluorescence units (RFU) in CE. Hovering the mouse over the bar in the graph yields the allele call and coverage, and a click will display the sequence. In addition to the sequence, it displays the SNPs in the upstream or downstream flanking region which are highly useful in increasing the discriminatory power of a DNA profile. Further down is the locus data in tabular form. There are three tabs including genotype, sequence histogram, and Integrative Genomics Viewer (IGV) link. IGV is a program that displays the sequencing data. The bases across the bottom represent the reference, and the bars reflect the data collected in the sequencing run. The colors reflect the forward and reverse strands. Stutter is reflected by 4 bp gaps in the sequence data. Base variants or mutations are shown with the one-letter representation of the base call. The RMP computation includes Y-STR markers.

The ThermoFisher Converge™ Forensic Analysis software merges CE and NGS data into one analysis. Converge is a case management, kinship, and paternity case tool. Users can review case status in the searchable Case Dashboard. The Dashboard lists the Case ID, Case Title, Creation Date, Owner and Priority, and the case overview, comments, and attachments such as crime scene photos and reports can be accessed by clicking on the case. A new case can be created with a title and identification number. A pop-up window enables the user to input the case status, notes, priority, and description and assign the case to an analyst. New subjects and details can also be created and associated with the same subject or case. Using the Upload Profiles tool, the user can upload CE and NGS data created with Applied Biosystems kits. Profiles using one or more CE kits and NGS are tabulated and combined to produce a composite profile. Empty cells with no allele call are colored red, and cells with calls from only one or two experiments are colored yellow. Loci not included in the kit are colored gray. STR allele calls using NGS can be displayed as graphs tiled five across on the screen. The graphs plot number of reads versus alleles. Read balance or imbalance can be evaluated at several loci simultaneously (Figure 5.11).

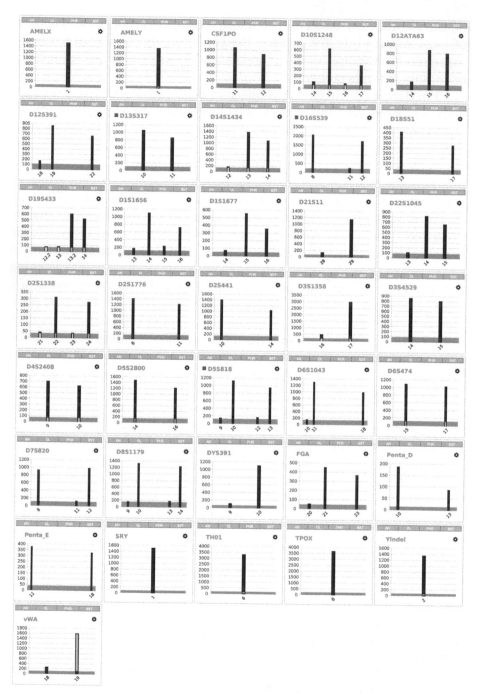

Figure 5.11 GlobalFiler NGS STR panel v2 data viewed with Converge software.

The allele calls are colored green in the graph, stutter peaks are colored light green, and peaks below the stochastic threshold are colored yellow. Flags at the top of the graph including allele number (AN), off-ladder alleles (OF), peak height ratio (PHR), and controlled concordance (BST) are green if they pass and red if they do not. Empty graphs signify loci with no data. The results at each locus can be probed individually. The allele calls and other peaks are color coded as in the graph, and the coverage and sequence are presented in a table. Expected alleles and off-ladder alleles are listed next to the graph of the locus results. The global parameters and STR thresholds are also shown to the right of the graph. The software indicates the analytical and interpretation thresholds in two different gray colors. All of the typed alleles and number of reads are shown in text and bar graph formats.

Case notes can be uploaded and viewed with DNA and other case data. The paternity portion incorporates tools for complex kinship analysis. The library type can be whole genome. For each barcode name, the sample, mapped reads, mean depth, and uniformity percentage value are displayed. The user can scroll down to HID_SNP_Genotyper_r94 and click the HID_SNP_Genotyper.html to view the results and output files by barcode name. For each barcode and sample name, the bases, ≥Q20 bases, reads, mean read length, read length histogram, and files to download the data (UBAM, BAM, BAI) are shown. Further down are more tabs including plugin summary, test fragments, chef summary, calibration report, analysis details, support contact, and software version. The user can download .fastq files and upload them for analysis into other apps or third-party applications.

For kinship and paternity analysis, users can drag and drop subjects and create pedigree trees on the platform's whiteboard. The trees can be tested using the null and alternative hypothesis. The user can adjust settings including minimum allele frequency computation strategy (e.g., FIVE_OVER_2N), fixed maximum allele frequency, mutation analysis model (e.g., MAM_TWO_PHASE), max mutation step, prior probability, population substructure, included ethnicity, conclusion ethnicity, and a box to check to calculate PI/RMNE. Incorrect subject linking detected by the software will be highlighted. The analysis results include combined LR, number of incompatible loci, number of loci excluded, conclusion ethnicity, prior probability, posterior probability, probability of exclusion (PE), and random man not excluded (RMNE). The software can process trio paternity, trio maternity, duo fatherless, and duo motherless cases. There is a conclusion box for the analyst to type in a conclusion. Electronic reports can be generated and signed electronically by inputting the username and password. The report can be downloaded and viewed in Microsoft Word.

After the data analysis steps are complete using Converge or the UAS, the data can be reported following the lab's SOP and data can be uploaded to NDIS and CODIS, if applicable. Data generated with the ForenSeq and the Precision ID kits have been approved to upload to CODIS.

5.4 Phenotype Analysis Using the Erasmus Server

The SNP loci contained in the HIrisPlex-S assay have been developed over years into the multiplex. The assay predicts hair color, eye color, and skin tone. If the HIrisPlex-S loci are not fully typed and not called by the UAS, the data can be analyzed at the Erasmus server if key loci are typed (e.g., rs 12913832). A few of the key loci can also lead to a prediction (Figure 5.12). Care must be taken to submit the typed alleles into Erasmus as the alleles vary on the top and bottom strands among loci in some kits (Table 5.2). The sequencing data for K562 were inputted into the Erasmus website (Figure 5.13), and the output was recorded (Figure 5.14). For comparison, Figure 5.15 shows the UAS prediction output for the same sample. From the data presented here, the human cell line K562 was likely derived from a woman with red hair, brown eyes, pale to intermediate skin tone and a European ancestry.

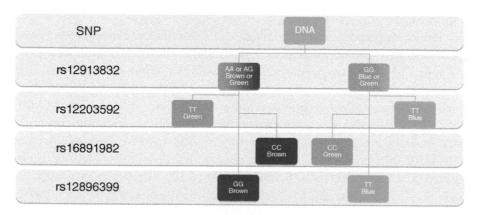

Figure 5.12 Eye color prediction tree using SNPs.

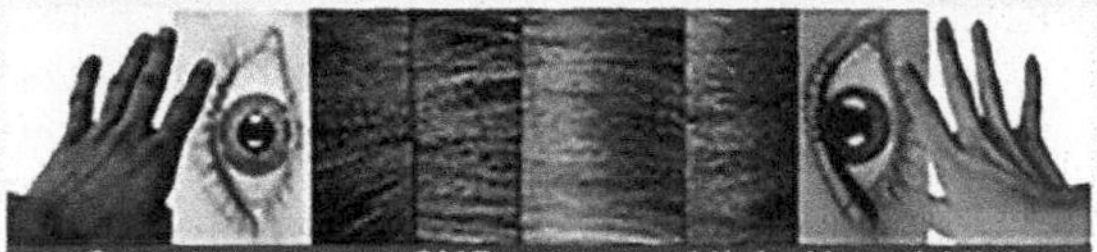

	Gene	SNP	Allele	No. of Alleles
1	*MC1R*	rs312262906	A	**0** 1 2 NA
2	*MC1R*	rs11547464	A	**0** 1 2 NA
3	*MC1R*	rs885479	T	**0** 1 2 NA
4	*MC1R*	rs1805008	T	**0** 1 2 NA
5	*MC1R*	rs1805005	T	**0** 1 2 NA
6	*MC1R*	rs1805006	A	**0** 1 2 NA
7	*MC1R*	rs1805007	T	0 **1** 2 NA
8	*TUBB3*	rs1805009	C	0 **1** 2 NA
9	*MC1R*	rs201326893	A	**0** 1 2 NA
10	*MC1R*	rs2228479	A	**0** 1 2 NA
11	*MC1R*	rs1110400	C	**0** 1 2 NA
12	*SLC45A2*	rs28777	C	**0** 1 2 NA
13	*SLC45A2*	rs16891982	C	**0** 1 2 NA
14	*KITLG*	rs12821256	G	**0** 1 2 NA
15	*LOC105374875*	rs4959270	A	**0** 1 2 NA
16	*IRF4*	rs12203592	T	**0** 1 2 NA
17	*TYR*	rs1042602	T	0 1 **2** NA
18	*OCA2*	rs1800407	A	**0** 1 2 NA
19	*SLC24A4*	rs2402130	G	**0** 1 2 NA
20	*HERC2*	rs12913832	T	0 **1** 2 NA
21	*PIGU*	rs2378249	C	**0** 1 2 NA
22	*LOC105370627*	rs12896399	T	**0** 1 2 NA
23	*TYR*	rs1393350	T	**0** 1 2 NA
24	*TYRP1*	rs683	G	**0** 1 2 NA
25	*ANKRD11*	rs3114908	T	0 1 2 NA
26	*OCA2*	rs1800414	C	0 1 2 NA
27	*BNC2*	rs10756819	G	0 1 2 **NA**
28	*HERC2*	rs2238289	C	0 1 2 **NA**
29	*SLC24A4*	rs17128291	C	0 1 2 **NA**
30	*HERC2*	rs6497292	C	0 1 2 **NA**
31	*HERC2*	rs1129038	G	0 1 2 **NA**
32	*HERC2*	rs1667394	C	0 1 2 **NA**
33	*TYR*	rs1126809	A	0 1 2 **NA**
34	*OCA2*	rs1470608	A	0 1 2 **NA**
35	*SLC24A5*	rs1426654	G	**0** 1 2 NA
36	*ASIP*	rs6119471	C	0 1 2 **NA**
37	*OCA2*	rs1545397	T	0 1 2 **NA**
38	*RALY*	rs6059655	T	0 1 2 **NA**
39	*OCA2*	rs12441727	A	0 1 2 **NA**
40	*MC1R*	rs3212355	A	0 1 2 **NA**
41	*DEF8*	rs8051733	C	0 1 2 **NA**

Display Predicted Phenotype   Download Predicted Phenotype

Figure 5.13 Erasmus SNP input for K562 prediction of phenotype.

Table 5.2 Key Notes for Inputting UAS Calls to Erasmus Server

ForenSeq/UAS iiSNPs called on opposite strand of HIrisPlex-S	rs 1042602
	rs 12821256
	rs 12913832
	rs 1393350
	rs 2378249
	rs 683
	rs 885479
Loci with alternative names in ForenSeq/UAS as HIrisPlex-S	N29insA = rs 312262906
	Y152OCH = rs 201326893

Loci courtesy of Adam Klavens.

Predicted phenotype		
	p-value	AUC Loss
blue eye	0.05	0
intermediate eye	0.114	0
brown eye	0.836	0
blond hair	0.026	0
brown hair	0.05	0
red hair	0.923	0
black hair	0.001	0
light hair	0.974	0
dark hair	0.026	0
very pale skin	0.075	0.021
pale skin	0.492	0.027
intermediate skin	0.429	0.02
dark skin	0.004	0.015
dark to black skin	0	0.002

Figure 5.14 Erasmus K562 phenotype prediction.

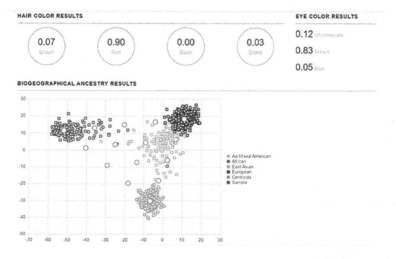

Figure 5.15 UAS biogeographical ancestry and phenotype prediction for K562 prepared with ForenSeq.

5.5 Other Sequence Analysis Software

NGS produces a lot of data that must be analyzed to extract all of the potential information from the samples. NGS raw data can be output in a variety of formats. Raw NGS data analysis from any of the library prep kits can be performed using a software pipeline.

All Illumina instruments output NGS data in the .bcl format which includes base calls per cycle and quality of each call. If samples are multiplexed, demultiplexing using the bcl2fastq program converts .bcl format to .fastq format. The .fastq format is the sequencing file format used by the bioinformatics community. It contains sequence data and quality information. It consists of four lines in each read. First, there is a sequence identifier (e.g., @SeqID), followed by the DNA sequence, then a plus symbol used as a spacer, and finally the Phred quality score (Q) (e.g., !"AAA***)%??5)))). With the fluorescent measurements, there is overlap between the colors and potential overlap of the bases and incorrect calls. The user is not shown the fastq file in UAS, but it is used to produce graphs and tables. FastQC can perform quality control on fastq files. It provides a summary report with visuals such as box and whisker plots. It can be used before and after other programs to evaluate the quality of the data. Raw fastq files generated using the PowerSeq 46GY System Prototype kit can be analyzed in STRait Razor v2.0 (Riman et al. 2020).

Raw sequence data can be analyzed and compared to other sequences including the sequence of a reference genome using sequence alignment. Sequence alignment enables the detection of variants. Analysis programs include BWA, Bowtie 2, Maq, Stampy, and Novoalign. Maq and Bowtie use the computational strategy called indexing to organize data using short sequences in the files. Maq uses spaced seed indexing. The read is divided into four segments of equal length called "seeds." Bowtie uses Burrows-Wheeler transform. It can analyze the full human genome using only 2 GB of memory, whereas Maq would require more than 50 GB of memory to analyze the sequences efficiently. If a reference sequence is not available, the sequences can be aligned *de novo* using programs including ABySS and SOAPdenovo. The sequences are compared and checked for overlap to build larger contiguous sequences called contigs until a complete contig is prepared for the entire genome of the organism. The file produced is called a sequence alignment map (SAM) file. The SAM file is the universal file for genomic sequence data. The sequence and quality scores of the mapped reads are contained in the file as well as the location of the reads in the genome. The compressed binary version of the SAM file is the BAM file. Picard is a command-line tool that can read SAM and BAM files.

Variants can be called using additional programs. The mapped data is compared to the reference genome to identify SNPs, SNVs, and INDELs. Genome Analysis Toolkit (GATK) and SAMtools mpileup are two major programs for variant calling. They use Bayesian algorithms to compare the sequences. The data is outputted in .vcf files. Data visualization tools include Integrative Genomic Viewer or the UCSC Genome Browser.

RNAseq analysis will differ from whole genome analysis as the reads will only map to the coding regions of the genome. TopHat and STAR programs handle reads split as splicing junctions. Exome-Seq analysis covers protein-coding genes. Probes are used that bind to these regions. As the majority of the regions that code for genes that cause diseases are found in the exome, this data is rich in answers to clinical questions. Since sequencing data and bioinformatics tools can be overwhelming to the casual user, preconfigured workflows for various analyses including microbial population detection using 16S metagenomics analysis (Chapter 8) is an area of effort. The analyses can be saved and edited to achieve the lab's specific reporting goals.

5.6 Additional Tools for Mixture Interpretation

While the UAS and Converge have flags and tools for mixture analysis, there are additional tools that could be employed. Mixture Ace is a tool for the analysis of NGS mixture data. The Parson ISFG format is verbose but easy for computers to manage. Fastq files are used to produce graphs that look like electropherograms and relate peak height ratios. Isoalleles are not stacked as they would be in CE. Families are color-coded and stutter is assigned the same color as the parent peak. The STR profile in question can be compared to a reference profile to view which loci are exact matches (orange) or included (yellow).

ArmedXpert™ is another tool for mixture deconvolution. An audit file is generated as a csv file with the marker, sequences and length included. Sequence errors in Illumina data are almost always the same length as the parent allele. Errors can occur in the allele and flanking regions and are identified by low reads. When a homozygous locus is sequenced, by far all of the reads will be homozygous but it is expected to observe approximately fifty sequencing errors per 50,000 reads, each in a different area, as the Illumina platform makes an error approximately once every 1000 nucleotide bases. These can be insertions or deletions and can be identified by alignment. The Promega PowerSeq kit (not NDIS approved at this writing) leads to reads from both directions which is useful in mixture analysis (van der Gaag et al.

2016). Verogen recommends using data of not less than 650 total locus reads and 10 read minimum. Mixture Ace can deconvolute microhaplotypes for mixture analysis and contributor ratio.

5.7 Other NGS Sequence Data Analysis Tools

In addition to the UAS and Converge software developed for forensic applications, there are many additional commercial and open source software tools available to analyze the massive quantities of sequence data produced with NGS. Although there are too many to cover in this book, we highlight a few below.

There are sequence alignment and presentation tools such as ExPasy suite of bioinformatics tools, ClustalW multiple sequence alignment, and MUltiple Sequence Comparison by Log-Expectation (MUSCLE) (a faster and more accurate replacement for ClustalW2).

Sequence diversity databases and STR search tools include STRbase 2.0 beta, NCBI Search / BioProject for STRs with Forensic kit annotations, STRscan, lobSTR, toaSTR, HipSTR, POPSeq, and STRSeq. PopSeq is a human STR sequence diversity database.

STRSeq is a National Center for Biotechnology Information (NCBI) tool that catalogues "sequence diversity at human identification Short Tandem Repeat loci" (Gettings et al. 2017).

The new STRbase 2.0 Beta (https://strbase-b.nist.gov/) hosted by NIST is more user friendly and searchable than the original format. Now, researchers can search and download information for variable STR alleles and other human markers including allele size ranges and sequence motifs. Much of the data for Table 5.1 is found in STRbase 2.0. The NCBI website has a feature by which users can search by kit such as "ForenSeq" to locate STR microsatellite target repeat sequences and Accession numbers. The miscellaneous features (misc_feature) include links to highlight the targets in various kits including ForenSeq, Precision ID, and PowerSeq 46GY in the genome sequence at the bottom of the page.

STRs are difficult to genotype due to the high mutability of the repetitive sequences and PCR stutter errors that result in alignment errors. There are many sequence handling tools aimed at STR analysis. STRScan is a standalone software tool that uses a greedy algorithm for targeted STR profiling in next generation sequencing data. STRScan (http://darwin.informatics.indiana.edu/str/) was tested on the whole genome sequencing data from Venter et al. (2001) and the 1000 Genomes Project published in *Nature*. The results showed that STRScan can profile 20% more STRs in the target set that are missed by lobSTR and STR-FM. STRScan is particularly useful for the

NGS-based targeted STR profiling, e.g., in genetic and human identity testing. lobSTR (http://lobstr.teamerlich.org/) is another tool for profiling STRs from high-throughput sequencing data that performed well with Y-STR data, and the toaSTR tool (https://www.toastr.de/) can be used to call STRs from massively parallel sequencing data independent of the instrument platform and the forensic kit used. HipSTR (Haplotype inference and phasing for Short Tandem Repeats) (https://hipstr-tool.github.io/HipSTR) was designed to perform profiling of heritable and *de novo* STR variations in genome data using a specialized hidden Markov model to align reads and phase STRs using phased SNPs (Willems et al. 2017).

Several tools have also been developed for SNP analysis. These include SNPedia (a SNP data finding tool), ALFRED (a SNP data finding tool), SNiPlay (a SNP graphics tool), WebLogo (a SNP graphics tool), SNPServer (a SNP discovery tool), SNPdetector (a SNP detection tool), QualitySNPng (a SNP detection and visualization tool) (http://www.bioinformatics.nl/QualitySNPng/), dbSNP (a SNP location and sequence region tool), and PredictSNP (a tool to predict SNP disease effect). Scientists can locate gene and variant information for SNPs, including those for eye color, at SNPedia. The WebLogo tool displays variations in sequence data using the size and stacking of the nucleotide base letters. SNPs in genomic data can be located using NCBI's dbSNP. The ALlele FREquency Database (ALFRED) (https://alfred.med.yale.edu) contains gene frequency data for human populations and offers graphics to plot allele frequencies worldwide. SNiPlay is another tool for displaying SNP data in pie charts and distance trees. SNPServer (http://hornbill.cspp.latrobe.edu.au/snpdiscovery.html) can be used to locate candidate SNPs. SNPdetector uses the template and primers to map the primers, locate SNPs and STRs and genotype SNPs (Zhang et al. 2005). PredictSNP (https://loschmidt.chemi.muni.cz/predictsnp/) can classify the effects of nucleotide substitutions on genetic sequence and amino acids coded for (Bendl et al. 2016). STRUCTURE can be used to analyze global human genome datasets and generate neighbor-joining Fst trees (Lawson et al. 2018).

5.8 NGS Validation and Applications

Several labs have tested and published the results of their implementation of the commercial NGS kits in the past few years. Jäger et al (2017) published the developmental validation of ForenSeq and interpretation using the UAS. Labs have conducted an internal validation of the Precision ID

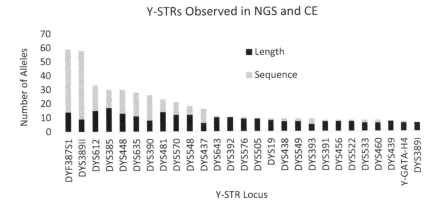

Figure 5.16 Number of alleles for Y-STRs analyzed using CE and NGS.

GlobalFiler™ NGS STR Panel amplification kit with the Ion Torrent S5™ sequencer (Faccinetto et al. 2019, Tao et al. 2019). Another lab evaluated the HID-Ion AmpliSeq™ Identity Panel using the Ion Torrent PGM™ platform (Guo et al. 2016) and the Illumina® ForenSeq™ DNA Signature Prep Kit using the MiSeq FGx™ (Guo et al. 2017). NIST reported upon sequence variation observed in single-source human DNA samples using the PowerSeq 46GY System Prototype kit with the Illumina sequencing platform (Riman et al. 2020). Becky Steffen and the Applied Genetics Group at NIST recently reported on the number of Y-STR alleles observed using CE and NGS at several loci (Figure 5.16); the sequence variations lead to approximately twice as many alleles at some loci.

The NGS kits contain the Y-STR loci typically found in supplementary kits and the aSTR and SNP loci (Figure 5.17). The analytical power of a 1 ng sample is greatly increased by NGS. Thus, NGS has been applied to case studies and compared to CE with difficult samples. For example, old blood samples from a Chinese Han population were typed revealing a high degree of polymorphism using MPS (Dai et al. 2019). Chemically compromised human remains from World War II era mass fatality events, including on the USS Oklahoma, the Battle of Tarawa, and the Cabanatuan Prison Camps, were analyzed using five DNA typing methods including mitotyping, autosomal STR typing using CE using two kits, Y-typing, and NGS (Edson et al. 2019). CE- and MPS-based DNA typing of petrosal bone and other skeletal remains was reported by two groups for forensic identification (Kulstein et al. 2018, Liu et al. 2020).

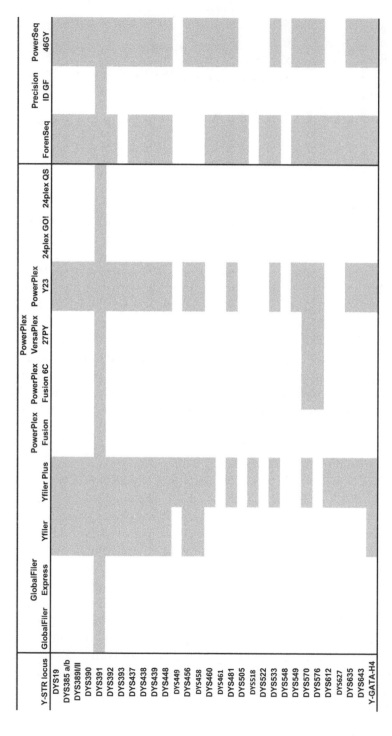

Figure 5.17 Number of alleles for Y STRs analyzed using CE and NGS.

Questions

1. List some tools that can be used for analyzing NGS data.
2. Are NGS reads analogous to fluorescence in CE? Explain.
3. How does the software algorithm locate the STR and SNP loci and assign alleles?
4. What does it mean if the human sequencing control fails?
5. How many reads are needed for high-quality NGS data?
6. What are some issues with NGS data that can complicate data analysis?
7. Explain how sequence polymorphisms can be used in human identification in cases of identical twins.
8. Explain how the prediction of biogeographical ancestry can be used in a case.
9. Explain how the prediction of eye color, hair color, and skin tone can be used in a case.
10. Explain the steps to performing random match probability calculations in the software.

References

Bendl, J., Musil, M., Stourac, J., Zendulka, J., Damborsky, J., and J. Brezovsky. "PredictSNP2: A unified platform for accurately evaluating SNP effects by exploiting the different characteristics of variants in distinct genomic regions." *PLoS Computational Biology* 12 (May 25, 2016): e1004962. doi:10.1371/journal.pcbi.1004962.

Dai, W., Pan, Y., Sun, X., Wu, R., Li, L., and D. Yang. "High polymorphism detected by massively parallel sequencing of autosomal STRs using old blood samples from a Chinese Han population." *Scientific Reports* 9, no. 1 (December 12, 2019): 18959. doi:10.1038/s41598-019-55282-9.

Edson, S.M. "The effect of chemical compromise on the recovery of DNA from skeletonized human remains: A study of three World War II era incidents recovered from tropical locations." *Forensic Science, Medicine, and Pathology* (November 12, 2019). doi:10.1007/s12024-019-00179-2.

Faccinetto, C., Serventi, P., Staiti, N., Gentile, F., and A. Marino. "Internal validation study of the next generation sequencing of Globalfiler™ PCR amplification kit for the Ion Torrent S5 sequencer." *Forensic Science International: Genetics Supplement Series* 7, no. 1 (December 2019): 336–338. doi:10.1016/j.fsigss.2019.10.002.

Gettings, K.B., Borsuk, L.A., Ballard, D., Bodner, M., Budowle, B., Devesse, L., King, J., Parson, W., Phillips, C., and P.M. Vallone. "STRSeq: A catalog of sequence diversity at human identification Short Tandem Repeat loci." *Forensic Science International: Genetics* 31 (November 2017): 111–117. doi:10.1016/j.fsigen.2017.08.017.

Guo, F., Zhou, Y., Song, H., Zhao, J., Shen, H., Zhao, B., Liu, F., and X. Jiang. "Next generation sequencing of SNPs using the HID-Ion AmpliSeq™ Identity Panel on the Ion Torrent PGM™ platform." *Forensic Science International: Genetics* 25 (November 2016): 73–84. doi:10.1016/j.fsigen.2016.07.021.

Guo, F., Yu, J., Zhang, L., and J. Li. "Massively parallel sequencing of forensic STRs and SNPs using the Illumina® ForenSeq™ DNA Signature Prep Kit on the MiSeq FGx™ Forensic Genomics System." *Forensic Science International: Genetics* 31 (November 2017): 135–148. doi:10.1016/j.fsigen.2017.09.003.

Jäger, A.C., Alvarez, M.L., Davis, C.P., Guzmán, E., Han, Y., Way, L., Walichiewicz, P., Silva, D., Pham, N., Caves, G., Bruand, J., Schlesinger, F., Pond, S.J.K., Varlaro, J., Stephens, K.M., and C.L. Holt. "Developmental validation of the MiSeq FGx Forensic Genomics System for Targeted Next Generation Sequencing in Forensic DNA Casework and Database Laboratories." *Forensic Science International: Genetics* 28 (May 2017): 52–70. doi:10.1016/j.fsigen.2017.01.011.

Kulstein, G., Hadrys, T., and P. Wiegand. "As solid as a rock-comparison of CE- and MPS-based analyses of the petrosal bone as a source of DNA for forensic identification of challenging cranial bones." *International Journal of Legal Medicine*, 132, no. 1 (2018): 13–24. doi:10.1007/s00414-017-1653-z.

Lawson, D.J., van Dorp, L., and D. Falush. "A tutorial on how not to over-interpret STRUCTURE and ADMIXTURE bar plots." *Nature Communications* 9 (August 14, 2018): 3258. doi:10.1038/s41467-018-05257-7.

Liu, F., van Duijn, K., Vingerling, J.R., Hofman, A., Uitterlinden, A.G, Janssens, A.C.J.W, and M.H. Kayser. "Eye color and the prediction of complex phenotypes from genotypes." *Current Biology* 19, no. 5 (March 10, 2009): R192–R193. doi:10.1016/j.cub.2009.01.027.

Liu, Z., Gao, L., Zhang, J., Fan, Q., Chen, M., Cheng, F., Li, W., Shi, L., Zhang, X., Zhang, J., Zhang, G., and J. Yan. "DNA typing from skeletal remains: a comparison between capillary electrophoresis and massively parallel sequencing platforms." *International Journal of Legal Medicine* 134, no. 6 (November 2020): 2029–2035. doi:10.1007/s00414-020-02327-8.

Nievergelt, C.M., Maihofer, A.X., Shekhtman, T., Libiger, L., Wang, X., Kidd, K.K., and J.R. Kidd. "Inference of human continental origin and admixture proportions using a highly discriminative ancestry informative 41-SNP panel." *Investigative Genetics* 4 (July 1, 2013): 13. doi:10.1186/2041-2223-4-13.

Riman, S., Iyer, H., Borsuk, L.A., and P.M. Vallone. "Understanding the characteristics of sequence-based single-source DNA profiles." *Forensic Science International: Genetics* 44 (January 2020): 102192. doi:10.1016/j.fsigen.2019.102192.

Sampson, J., Kidd, K.K., Kidd, J.R., and H. Zhao. "Selecting SNPs to identify ancestry." *Annals of Human Genetics* 75, no. 4 (July 2011): 539–553. doi:10.1111/j.1469-1809.2011.00656.x.

Tao, R., Qi, W., Chen, C., Zhang, J., Yang, Z., Song, W., Zhang, S., and C. Li. "Pilot study for forensic evaluations of the Precision ID GlobalFiler™ NGS STR Panel v2 with the Ion S5™ system." *Forensic Science International: Genetics* 43 (November 2019): 102147. doi:10.1016/j.fsigen.2019.102147.

The 1000 Genomes Project Consortium. "An integrated map of genetic variation from 1,092 human genomes." *Nature* 491 (November 1, 2012): 56–65. doi:10.1038/nature11632.

The 1000 Genomes Project Consortium. "A global reference for human genetic variation." *Nature* 526 (October 1, 2015): 68–74. doi:10.1038/nature15393.

van der Gaag, K.J., de Leeuw, R.H., Hoogenboom, J., Patel, J., Storts, D.R., Laros, J., and P. de Knijff. "Massively parallel sequencing of short tandem repeats-population data and mixture analysis results for the PowerSeq™ system." *Forensic Science International: Genetics* 24 (September 2016): 86–96. doi:10.1016/j. fsigen.2016.05.016.

Venter, C.J., Adams, M.D., Myers, E.W., Li, P.W., Mural, R.J., Sutton, G.G., Smith, H.O., Yandell, M., Evans, C.A., Holt, R.A., and J.D. Gocayne. "The sequence of the human genome." *Science* 291, no. 5507 (February 16, 2001): 1304–1351. doi:10.1126/science.1058040.

Walsh, S., Lui, F., Ballantyne, K.N., van Oven, M., Lao, O., and M. Kayser. "IrisPlex: A sensitive DNA tool for accurate prediction of blue and brown eye colour in the absence of ancestry information." *Forensic Science International: Genetics* 5, no. 3 (June 2011): 170–180. doi:10.1016/j.fsigen.2010.02.004.

Walsh, S., Chaitanya, L., Clarisse, L., Wirken, L., Draus-Barini, J., Kovatsi, L., Maeda, H., Ishikawa, T., Sijen, T., de Knijff, P., Branicki, W., Liu, F., and M. Kayser. "Developmental validation of the HIrisPlex system: DNA-based eye and hair colour prediction for forensic and anthropological usage." *Forensic Science International: Genetics* 9 (March 2014): 150–161. doi:10.1016/j.fsigen.2013.12.006.

Willems, T., Zielinski, D., Yuan, J., Gordon, A., Gymrek, M., and Y. Erlich. "Genome-wide profiling of heritable and de novo STR variations." *Nature Methods* 14 (April 24, 2017): 590–592. doi:10.1038/nmeth.4267.

Zhang, J., Wheeler, D.A., Yakub, I., Wei, S., Sood, R., Rowe, W., Liu, P.P., Gibbs, R.A., and K.H. Buetow "SNPdetector: A software tool for sensitive and accurate SNP detection." *PLoS Computational Biology* (October 28, 2005). doi:10.1371/ journal.pcbi.0010053.

Next Generation Sequencing Troubleshooting

6

6.1 Troubleshooting NGS Sequencing

Following the sequencing run, the next generation sequencing (NGS) data can be interpreted and sufficiently complete profiles can be uploaded to the National DNA Index System (NDIS). As with CE, issues may arise from time to time with NGS. There are several points in the protocol and process in which errors or issues can arise and result in a partial profile or none at all. These include kit storage conditions, thermal cycler ramping rate or performance, processing time for normalization steps, and room temperature for library preparation and sequencing. Issues can also arise from analyst errors or deviating from manufacturer or validated protocols, leaving streaks on the flow cell, improper instrument settings, mechanical or software failure, or from using expired kits or recalled lot numbers. Additional issues may occur in the manufacturing process and may be determined by lot numbers. Trace and low-quality samples due to degradation, impurities, or random breakage in the DNA can all lead to problems in sequencing. Data transfers may be delayed if the instrument software loses connectivity with the instrument. Processing samples with a different PCR-based STR typing kit followed by CE can help resolve an issue with the samples or NGS kit or process.

6.2 Troubleshooting MiSeq FGx Instrument Failure

The Illumina MiSeq has demonstrated consistently to be a reliable and robust instrument which supports Verogen's decision to adopt the MiSeq and upgrade it to the MiSeq FGx instrument for forensic applications. However, as with all machines with mechanical parts, failures can and do happen, even if they are rare. Troubleshooting some of these possible issues will be the focus of this section.

When upgrading an existing MiSeq instrument to a MiSeq FGx, the software update will add a new screen from which a user can select research use only (RUO) or forensic application. Each existing user needs to create a new username and password in the Verogen Universal Analysis Software (UAS) to login, even if using the instrument in RUO mode. The new server must remain connected for the instrument to be able to run. One issue is

DOI: 10.4324/9781003196464-6   87

the server can lose connectivity; the problem can be addressed by rebooting the server. If the instrument loses the connection while data is being transferred to BaseSpace, it must also be restarted to initiate the transfer. Data from sequencing runs using ForenSeq is automatically transferred to the UAS. Users can initialize the instrument manually using the MiSeq Test Software by pressing the spinning arrows icon or shutdown the instrument, waiting a minute and restarting the instrument and again trying to initialize the instrument. Another issue that may occur is the MiSeq FGx Y-stage fails to return to the home position. The user can attempt to return the Y-stage to the home position via the MiSeq Test Software by selecting the icon with gears (Figure 6.1). Unfortunately, if the issue persists through the MiSeq Test Software, Verogen should be consulted for replacement by a field engineer. If the sequencing run fails to start, the user can wait a minute and restart and try to start the run again. Similarly, if the MiSeq FGx instrument gets stuck at a screen position during the wash step and the stop button does not stop the wash, the instrument will need to be rebooted by toggling the switch to off, waiting a minute, and toggling it to the on position. After rebooting the instrument, a maintenance wash should be performed. When loading consumables, the barcode may not be found or the user may have inserted an expired consumable. Replacing the consumable or manually overriding the warning will allow the user to initiate the run. After the sequencing is completed, the status of the HSC or PhiX control (RUO mode) should be evaluated to determine if they passed/failed. Failure of the controls could indicate an issue with the integrity of the control itself or a problem with the library preparation process. The sequencing run can be repeated or the library prep

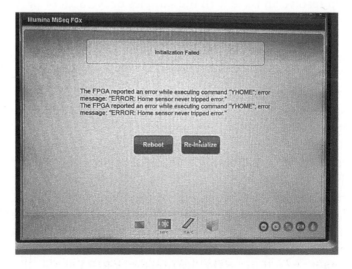

Figure 6.1 MiSeq FGx Y-stage home error.

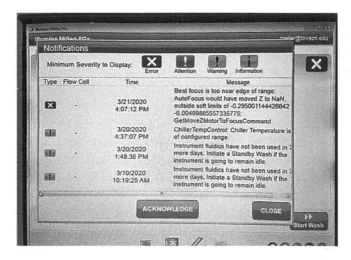

Figure 6.2 MiSeq FGx camera focus error.

and sequencing can be repeated. If the HSC/PhiX control passes, but the phasing/prephasing/clustering flags are raised, the user can repeat the library preparation steps with high quantity input DNA and perform the bead-based steps more quickly with fewer samples in a batch to reduce primer dimers and increase sample target amplicons. If the problem continues after above is performed, maintenance should be performed to refocus the camera collecting the raw fluorescence data (Figure 6.2).

6.3 Troubleshooting MiSeq FGx Run Failure

Several problems can lead to issues and orange or red flags in the Verogen MiSeq FGx sequencing run (ForenSeq™ DNA Signature Prep Reference Guide). If a low cluster density or high cluster density with a low cluster passing filter % is detected, an issue with library normalization or quantity of input DNA is indicated. Adding too much or low quantity of input DNA, an incorrect volume of beads in the library normalization step or processing the samples too slowly during the bead-based steps may result in the analyst needing to repeat the library preparation steps and the sequencing. The flags may result from issues with PCR ramp failure or improper temperature program. Short or no amplicons and primer dimers indicate that the library preparation steps need to be redone with the correct PCR ramp settings and/or a working thermocycler. A low cluster density may indicate too little DNA input, a failure in the library prep steps, or that the diluted library was insufficiently heated prior to loading to the cartridge for sequencing.

These issues were previously discussed in the library preparation section of Chapter 3 and can be diagnosed using an agarose or polyacrylamide gel or using a Bioanalyzer or QIAxcel instrument. Issues with high phasing can be caused by environmental issues such as the room temperature being too high. The instrument run files can be helpful in troubleshooting (e.g., D:\ Illumina Maintenance Logs->Temperature Log Chiller or ->Temperature Log flow cell). High prephasing often indicates the need for the instrument to undergo a maintenance wash. Best practices include performing sufficient wash steps prior to each sequencing run and refilling the wash tray and bottle after every wash using fresh Tween 20 and bleach wash solutions. Reads produced with sequence runs that produce orange passing quality metrics often produce results that are sufficient to be used for analysis. However, this could indicate that the library preparation steps need to be repeated and the samples need to be resequenced, or that a new flow cell and cartridge should be used to resequence the samples.

To obtain the best possible profile for each sample, it is very important to follow the protocol as instructed by the manufacturer for optimal performance of the ForenSeq kit. Errors in these steps will result in poor sequencing runs. To avoid contamination, aerosol-resistant, filter tips should be used and changed between reagents and samples. The no template control (NTC) should be free of allele calls. Reagents should be mixed well but the master mix should not be vortexed. Plates should be sealed with film and mixed following the protocol. For example, for optimal PCR performance, reaction components must be well mixed in the liquid and sealed tightly to avoid evaporation. If the PCR amplification steps did not work properly and adaptors are added but the targets are not sufficiently enriched, adapter dimers can form in the library normalization step and be carried through to sequencing at an unusually high proportion to the amplified product. If adaptor dimers are formed as determined by sizing, the sample can be repurified or reprocessed though library prep. Some protocol deviations may result from laboratory limitations and existing tools and ancillary instruments. For example, the ForenSeq library preparation steps will still succeed if a shaker that is limited to 1500 rpm is substituted for one that performs up to 1800 rpm. Similarly, setting a 4% ramp on the Veriti thermocycler is not optimal but will not cause the amplification to fail entirely. In contrast, an issue with the kit components or the PCR 1 step can cause PCR 1 to fail. If PCR 2 is successful, the adaptors and indexes are added to the forward and reverse tags. In the sequencing run, the quality metrics may all pass, but the only data would be of the sequenced adaptors because the target was not successfully amplified in PCR 1, so no alleles will be called. It is extremely important to check the amplicon sizes after PCR 1 and PCR 2 using a diagnostic gel, Bioanalyzer, tape station, or fragment analyzer with a suitable ladder to check that the process is proceeding normally to avoid unnecessary time and cost in proceeding with steps that will fail. Correctly amplified targets will be approximately 800 bp. Small base pair fragments indicate an issue to be examined.

As previously described, working quickly through the bead-based steps for purification and normalization is essential for optimal performance. Thus, it is best to work with a modest number of samples when performing library preparation. The magnetic beads should be warmed to room temperature prior to use. When working with the magnetic beads for the library purification and normalization steps, multichannel pipettes should be used to ensure consistent volumes and quick processing of samples through these steps. The magnetic beads settle quickly and should be mixed well to resuspend the beads prior to pipetting and pipettes should be checked to see if they are drawing up equal quantities of beads. When the samples are mixed with the beads, they should be mixed thoroughly to ensure complete binding. The supernatant should be drawn off slowly to avoid sample loss by pipetting the DNA-bound beads. The beads should be washed with freshly prepared 80% ethanol. The ethanol should be removed completely from the beads and care should be taken to avoid drawing up the beads. However, care must be taken so that the beads do not dry out between the ethanol washes and DNA elution steps.

To address the run failure problem further, the data that can be recovered from a failed sequencing run will vary depending upon when the run failed. If Read 1 completes but Index 1, Index 2, and Read 2 fail, the data for Read 1 will be lost because the samples are multiplexed and, without the index reads, cannot be demultiplexed (Figure 6.3). However, if Read 1, Index 1, and Index

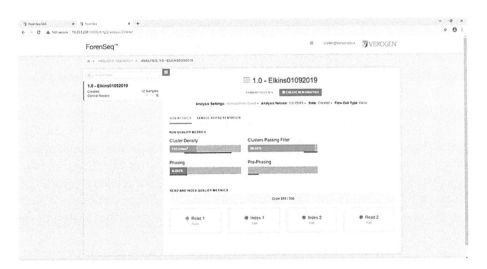

Figure 6.3 MiSeq FGx run failure viewed in UAS.

2 complete normally, but Read 2 fails, the Read 1 data can be demultiplexed and assigned. The run will lack the dual confirmation from Read 2. A low number of reads or cluster density overall may indicate a flow cell issue with a poor-quality oligo lawn, low template DNA input, or processing issues with the bead steps during library preparation.

Expectations should be calibrated based on the quality and quantity of the samples. As with the DNA standard tested with the samples, high-quality samples with at least 1 ng of DNA in 5 μL added to PCR1 should be expected to yield a full profile. The full nanogram of input DNA should be used, if possible. Samples can be concentrated to achieve the recommended 0.2 ng/μL input concentration. Standards including 2800M and the human sequencing control (HSC) should pass all checks, and a full profile should be achieved with the 2800M to have confidence in the data quality for the rest of the samples. The NTC should have no allele calls. Female samples should be devoid of detectable Y markers. If the MiSeq FGx quality flags turn orange or red, the sequencing run may have to be repeated. However, samples known to be degraded or contain PCR inhibitors as indicated from the quantification step should not be expected to yield full profiles using a traditional CE DNA typing method or NGS. Adding more DNA input may be impossible based upon the evidence sample and DNA yield. Low template samples may not yield full profiles, and a consensus profile may need to be determined from several runs (Table 6.1).

6.4 Troubleshooting Ion Series Run Failure

The mechanical parts and consumables are all subject to failure or quality assurance (QA) issues. As with the MiSeq FGx, the ThermoFisher Ion series instruments have moving parts that can get stuck or fail to initialize. Manufacturing plants can experience QA issues. The aforementioned thermocycler failures, analyst errors, instrument failures, and consumable issues can occur with a kit from any manufacturer.

If upon analyzing data from a sequencing run, a base variant is encountered that has not previously been reported or the NGS data is questioned or questionable for any reason, Sanger sequencing with the ThermoFisher BigDye Direct sequencing kit can be used to confirm sequence variants identified in NGS runs. In the Ion Reporter software, the user can select "Order CE primers" or use the ThermoFisher Primer Designer web tool to design the

Table 6.1   Troubleshooting the MiSeq FGx Instrument and Sequencing Runs

Problem	Solution
PCR ramp failure or improper temperature program	Redo library preparation steps with the correct ramp settings and working thermocycler
MiSeq FGx instrument does not initialize	Run MiSeq Test Software by pressing the gears icon or shutdown, wait a minute and restart and retry
MiSeq FGx Y-stage does not return to home	Run MiSeq Test Software by pressing the spinning arrows icon and call Verogen for maintenance if the software does not bring it to the home position
Sequencing run is not starting	Shutdown instrument, wait a minute and restart and retry
Instrument is not connecting to server	Reboot server
Barcode not found or expired consumable	Replace or manually override warning
Run failed during sequencing	Check HSC or PhiX control (RUO) to determine if they pass/fail If the HSC/PhiX control passes, redo library preparation steps with high quantity input and perform bead-based steps more quickly with fewer samples in a batch to reduce primer dimers If the problem continues, maintenance should be performed to refocus the camera
Low cluster density – orange or red flag	Increase DNA input quantity and redo library preparation or retry with carefully heating samples prior to loading them onto the cartridge or rerun with a new flow cell to address microfluidics issues
High phasing – orange or red flag	Check room temperature and reduce if needed
High prephasing – orange or red flag	Perform maintenance wash with freshly-prepared solutions
High phasing, prephasing, or low cluster density	Redo library preparation

PCR primers for the loci of interest. The PCR primers need to be designed with M13 tails. The Next Generation Sequencing Confirmation (NGC) module in the cloud can be used to compare the Sanger sequencing and NGS results using the PCR primers or Assay ID and the .vcf file, respectively (Precision ID GlobalFiler™ NGS STR Panel v2 with the HID Ion S5™/HID Ion GeneStudio™ S5 System Application Guide).

Questions

1. List issues that may arise during the sequencing run on the instrument and cause it to fail.
2. List problems that can occur if the sample residence time is too long during the bead-based normalization steps.
3. List outcomes that may be observed with incorrect PCR ramp, temperature, and hold settings.

References

ForenSeq™ DNA Signature Prep Reference Guide. August 2020. Accessed May 21, 2021. https://verogen.com/wp-content/uploads/2020/08/forenseq-dna-signature-prep-reference-guide-VD2018005-c.pdf.

Precision ID GlobalFiler™ NGS STR Panel v2 with the HID Ion S5™/HID Ion GeneStudio™ S5 System Application Guide. Revision 15 November 2018. Accessed May 21, 2021. https://assets.thermofisher.com/TFS-Assets/LSG/manuals/MAN0016129_PrecisionIDSTRIonS5_UG.pdf.

Mitochondrial DNA Typing Using Next Generation Sequencing

7

7.1 Introduction to Mitochondrial DNA Typing

Although crime labs primarily collect short tandem repeat (STR) nuclear DNA profiles for forensic DNA typing because they allow for superior statistical analysis and human differentiation, mitochondrial DNA (mtDNA) profiles provide forensic scientists with a valuable tool for identifying maternal lineages, performing genetic genealogy, and identifying individuals when the DNA recovered is damaged, degraded, or too low in quantity for STR analysis (Wallace et al. 1999, Eduardoff et al. 2017, Rathbun et al. 2017, Strobl et al. 2019). Mitochondrial DNA typing is valuable for analyzing challenging samples and providing investigative leads in missing persons cases (Cuenca et al. 2020), mass disaster cases (Biesecker et al. 2005), cold cases, and historic investigations (Hickman et al. 2018, Buś et al. 2019, Ambers et al. 2020); evaluating mother-child pairs (Ma et al. 2018); and differentiating monozygotic twins (Wang, Zhang et al. 2015, Wang, Zhu et al. 2015). Mitochondrial DNA typing has been used to analyze hair shafts, bone, teeth, and degraded or compromised samples, and to analyze DNA adhering to an earphone (Ivanov et al. 1996, Holland and Parsons, 1999, Seo et al. 2002, Remualdo and Oliveira 2007, Chaitanya et al. 2015, Lee et al. 2015, Parson et al. 2015, Marshall et al. 2017, Gallimore et al. 2018, Gaag et al. 2020, Kim et al. 2020).

Mitochondrial DNA typing has been performed in cases since 1996 (Ivanov et al. 1996). The mitochondrion is an unusual organelle that possesses its own genome outside of the cell's nucleus. Its genome is comprised of 16,569 bp (Anderson et al. 1981). Its small circular structure means it often remains intact when nuclear DNA is degraded. Human egg and sperm cells contain mitochondria; however, upon fertilizing an egg, the sperm mitochondria are degraded by an endonuclease (Chan and Schon 2012). While there is only one copy of the nuclear DNA genome in each cell, there can be as many as one to fifteen copies of the mitochondrial chromosome per mitochondria, up to 1000 mitochondria per cell, and therefore as many as hundreds to thousands of copies of the mitochondrial genome in each cell (Budowle et al. 2003). A DNA sample with no detectable copies of genomic DNA can have over 20,000 copies of mitochondrial DNA (Parson et al. 2015). Mitochondrial DNA typing can be used to determine if bones and teeth derive from the same or a different skeleton or family. Thus, the high copy

DOI: 10.4324/9781003196464-7   95

number of mitochondrial chromosomes in cells makes them ideal for genetic analysis and use as a matrilineal screening tool.

7.2 The Sequence of the Mitochondrial Chromosome

Routine sequencing or variable position typing of mtDNA has enabled its use in forensic DNA screening assays and haplotype determination. The mitochondrial chromosome originally sequenced in 1981 is referred to as the Anderson sequence or Cambridge Reference Sequence (CRS) (Anderson et al. 1981). With improved DNA sequencing technologies, the same sample was resequenced in 1999 and is referred to as the revised CRS (rCRS) (NCBI NC_012920) (Andrews et al. 1999). Resequencing the CRS addressed technical errors that had been flagged over the years through public inquiry. Base changes in the CRS were identified by position number. The variations between the CRS and rCRS are listed in Table 7.1.

The mitochondrial chromosome is largely conserved as it contains genes that code for proteins essential for metabolism; mutations can lead to some diseases. It includes thirty-seven genes that encodes twenty-two tRNAs, two rRNAs, thirteen proteins involved in oxidative phosphorylation, and a 1122 bp "control" region of non-coding DNA (Butler 2005). The control region is not known to code for any medically or phenotypically significant genes. The mitochondrial chromosome does not contain STRs for use in identity typing. However, variation in the mitochondrial genome has been studied extensively. The control region has been found to contain many SNPs, and

Table 7.1 Variations between the CRS and rCRS Mitochondrial Chromosome Sequences

Base Position	CRS	rCRS	Comment
311–315	CCCCC	CCCC	5C instead of more common 6C
3106–3107	CC	C	sequencing error, 3107del*, gap denoted by N
3423	G	T	sequencing error
4985	G	A	sequencing error
9559	G	C	sequencing error
11,335	T	C	sequencing error
13,702	G	C	sequencing error
14,199	G	T	sequencing error
14,272	G	C	error due to bovine DNA
14,365	G	C	error due to bovine DNA
14,368	G	C	sequencing error
14,766	T	C	error due to HeLa DNA

Source: Data from Andrews et al. (1999).

additional SNPs have been found scattered in the base sequence between the coding genes. The SNP variations have found uses in matrilineal typing and to distinguish individuals (Coble et al. 2004, Warner et al. 2006, Fridman and Gonzalez 2009, Holland et al. 2018). Mitochondrial DNA typing can be used to determine if bones and teeth derive from the same or a differ ent skeleton, family relationships, and maternal relatives in genetic geneal ogy (Bruijns et al. 2018). Mitochondrial DNA typing was used to assign the remains interred in the Tomb of the Unknown Soldier at Arlington National Cemetery to First Lieutenant Blassie who served in the Vietnam War and died in 1972 (Butler 2005). Within the control region, there are three regions that have been found to contain the most variation and are referred to as the highly or hyper variable (HV) regions I, II and III. (Fridman and Gonzalez 2009). The most common variations are shown in Table 7.2. Data from HV1 and HVII are most often used to differentiate individuals and families, with HVIII used to resolve indistinguishable HV1/HV2 samples (Budowle et al. 1999, Bini et al. 2003, Fridman and Gonzalez 2009). Other SNPs are located between and around these three hypervariable regions. As mtDNA mutation rates are ten to twenty times higher than nuclear DNA genes due to the low fidelity of the mtDNA polymerase and its lack of repair mechanisms, it can be used as an ancestral clock (Wallace et al. 1999, Budowle et al. 2003). The aver age nucleotide variation in these regions is estimated at 1.7% so variations are clocked in only a few generations.

The frequencies of variation at these sites has been studied extensively and are cataloged in the MITOMAP database (MITOMAP). MITOMAP contains a compilation of mtDNA SNP data from diverse populations worldwide and includes the observation frequency of SNPs. A mtDNA

Table 7.2 Frequently Probed Mitochondrial DNA SNP Positions in the Variable Regions (HVI, HVII, and HVIII) and Outside the Control Region (Other)

HVI (16,024–16,365)	HVII (73–340)	HVIII (438–574)	Other
16,051	73	477	3010
16,093	146	489	4580
16,126	150		4793
16,129	152		5004
16,223	189		7028
16,270	195		7202
16,278	198		10,211
16,304	200		12,858
16,309	247		14,470
16,311	310		16,519
16,362			

mutation occurs approximately once every 8000 years. An mtDNA haplogroup is defined by differences in the mtDNA sequence. A haplogroup may vary by only one SNP from another haplogroup and are named with letters from A to Z in order of their discovery. The International HapMap Project seeks to develop a haplotype map of the human genome and can be used to find genetic variations implicated in disease and geographic genetic origins.

7.3 Mitochondrial DNA Typing Methods

Prior to NGS, mtDNA SNPs were detected using one or more of several available assays including a primer extension assay (Vallone et al. 2004), mini-sequencing assays (Gabriel et al. 2001), SNaPshot assays (Quintáns et al. 2004), denaturing high performance liquid chromatography (LaBerge et al. 2003), a three-dye fluorescence labeling mitochondrial-SNP kit called Expressmarker mtDNA-SNP60 (Zhang et al. 2018), and polymerase chain reaction (PCR) high resolution melt (HRM) analysis (Dobrowolski et al. 2009, Elkins 2013). A drawback of these methods is the limited sequence information and discriminating power. Depending upon the approach and the SNPs of interest, mtDNA typing could take a few hours to a couple weeks using these methods. Human mtDNA standard reference material (SRM) 2392 and SRM 2392-I have been produced by the National Institute of Standards and Technology (NIST) for use a positive control in amplification and sequencing (Levin et al. 2003).

7.4 Mitochondrial DNA Typing Using Next Generation Sequencing

NGS using kits and methods for control region and whole mitochondrial chromosome sequencing offer more discrimination power than previous methods. Additionally, applying NGS to mtDNA forensic DNA typing for bone, teeth, and hair human remains and degraded DNA samples for which only partial or no STR profile was recovered allows forensic scientists to obtain some genetic information on these samples. For example, Ambers et al. (2006) applied NGS to human remains found in Deadwood, South Dakota and determined the skeleton to be of European (Caucasian) ancestry, concordant with the anthropological findings. There are several commercial and published approaches for mitochondrial DNA typing using NGS including the AFDIL method (Fendt et al. 2009), the ForenSeq mtDNA Control Region Kit (Verogen), ForenSeq mtDNA Whole Genome kit (Verogen), the

QIAseq Human Mitochondria Panel (Qiagen), QIAseq Investigator Human Mitochondria Control Region Panel (Qiagen), Precision ID mtDNA Control Region Panel (ThermoFisher Scientific), and the Precision ID mtDNA Whole Genome Panel (ThermoFisher Scientific). The commercial library preparation kits include all of the required multiplexed primer sets, PCR reaction mixes, control standards, indexes, purification reagents, normalization beads, and buffers. AFDIL's published approach details their method and researchers can obtain all of the required library preparation primer sequences, reagents, and consumables from suppliers and replicate the process in their labs at low cost (Fendt et al. 2009). SRM 2392 consists of three mitochondrial genome samples. SRM 2392-I consists of the HL-60 cell line. These standards have been fully typed. Control standard samples are used to determine concordance of the NGS results with sequence data produced by Sanger sequencing, and preliminary evaluations of the commercial kits have demonstrated them to be accurate in their performance.

In cases in which samples are damaged, old or compromised, little DNA may be recovered and it may be of low quality. Scientists can frequently obtain mtDNA profiles from less than a picogram of total DNA. Working with such low quantities of DNA requires a very clean lab and commitment to contamination elimination measures. Prior to proceeding to library preparation, the co-extracted nuclear DNA may need to be digested to avoid interference with the mtDNA assay primer set target amplification. DNase I selectively digests nuclear DNA. Restriction enzymes, such as HaeIII, CfoI, or MspI, that selectively digest GC-rich (nuclear) DNA can be used. Relatively large fragments of multi-copy mtDNA result. Alternatively, for degraded and low-quantity templates and as an alternative to DNA digestion, whole genome amplification (WGA) can be performed prior to library prep to enrich the mtDNA targets. One such kit is the Qiagen REPLI-g Mitochondrial DNA kit. REPLI-g can be used to amplify human and non-human mtDNA and increase the sensitivity of NGS.

The ForenSeq mtDNA Control Region Kit targets the control region range 16008–594 (16008–16569 and 1–594) (Walichiewicz et al., 2019). The kit employs two primer sets of 122 primers to generate 18 primary amplicons less than 150 bp in length spanning and overlapping in the mtDNA control region where the majority of the variation is located and is based on research performed by McElhoe et al. (2014). To ensure there are no gaps in the sequence when aligning the sequences using bioinformatics, all of the amplicons overlap by ≥3 bp. If desired, custom primers can also be integrated in the platform. The recommended DNA input is 50 pg for each initial PCR reaction or a total input of 100 pg of gDNA per sample. Successful profiles have been demonstrated with a 12 μL maximum input of DNA extract from teeth, bone, or buccal cells and 0.5 cm of hair shaft (Gallimore et al. 2018).

The included positive control is HL-60. As with the ForenSeq Signature
Prep kit described in Chapter 3, the library preparation steps amplify and
enrich the target regions and add the forward and reverse tags and the adaptor and index sequences. PCR1 leads to enrichment of the targets and addition of the forward and reverse tags in two reactions for each of the samples.
During PCR2, the PCR1 products are pooled, and the i5 and i7 adaptors and
indexes are added. The average primary amplicon size is 118 bp, while some
are as small as 61 bp and the largest amplicon is 458 bp. The short amplicons are intended to lead to optimal results when amplifying degraded DNA.
To check the quantity, size, and quality of the libraries, the samples can be
analyzed using an agarose or polyacrylamide gel or using the Agilent DNA
1000 kit using the Agilent 2100 Bioanalyzer system. Following the PCR1
and PCR2 library preparation steps, either of two methods can be used for
library normalization: a bead-based normalization (BBN) and a fluorimetric
quantification-based normalization. Normalization is performed to achieve
more equal cluster densities on the flow cell and therefore better detection
of each sample upon pooling. Unfortunately, even though all or most of the
low-quantity samples will bind the beads, normalizing the high-quantity samples with the low-quantity samples can lead to a lowered cluster density of all
samples. The beads have a maximum binding capacity, and the high quantity
will be reduced when the beads reach their binding capacity. The bead-based
method is the standard method for high-throughput and can be automated.
The quantification-based method is low-throughput but can improve sample
representation for low input samples (<20 pg), which result in lower overall
library yield. In a recent study, smaller number of samples in a run, manual
quantitation and normalization led to four times greater coverage per sample
as compared to BBN with no adverse effects to read calling (ISHI poster). In
the bead-based normalization procedure, the prescribed quantity of beads
are added to each sample. The bead-based normalization can be performed
using the beads included in the ForenSeq kit or a different kit such as the
Nextera® XT DNA Sample Preparation kit can be used. The Verogen method
employs Illumina technology reaction mix cartridges and flow cells, and the
DNA fragments are sequenced using the Illumina MiSeq instrument that has
been upgraded to the Verogen MiSeq FGx model. According to the manufacturer, up to 48 mtDNA D-loop HV samples can be multiplexed in a sequencing run on a MiSeq Micro Reagent Kit v3 flow cell (2×151 cycle run with dual
index reads) and >48 can be sequenced on a standard flow cell. The amplicon start-end locations are 29–285, 172–408, 16997–16236, and 16159–16401.
The mtDNA whole genome samples are sequenced using a 2×251 cycle run
with dual index reads (amplicon start-end: 9397–1892 and 15195–9796). The
samples and indexes are assigned in the UAS. Each sequencing run completes
in approximately 18 hours.

There are two mtDNA NGS options available from ThermoFisher: the Precision ID mtDNA Whole Genome Panel kit and the Precision ID mtDNA Control Region Panel kit. Both are available in 48 or 96 sample options. The optimal input DNA for the Precision ID mtDNA Whole Genome Panel kit is 125 pg but as little as 2 pg of DNA can be used for it and the control region kit. The whole genome panel consists of 2 primer pools of 81 primers each. The mtDNA Whole Genome Panel kit employs a tiling approach employing 162 primer sets in two reactions and yields amplicons of only 163 bp, on average, in length with an average overlap of 11 bp. The Precision ID mtDNA Control Region Panel kit uses the same tiling approach with an average amplicon length of 153 bp and an average overlap of 18 bp. The control region kit targets the 1.2-kb control region (16024-576) which encompasses the HVI, HVII, and HVIII regions rather than the full 16,569 bp genome. NGS chip preparation can be performed on the Ion Chef or prepared manually and sequenced on an Ion GeneStudio S5 System instrument. When using the Ion Chef, there are only five pipetting steps and forty-five minutes of hands-on time as the rest is performed by the robot. On an Ion 510 Chip, thirty-seven samples can be run prepared using the mtDNA control region library. On an Ion 520 chip, fifty-six samples can be run simultaneously. When working with a mtDNA whole genome library, twenty-five samples can be run on the Ion 520 chip and thirty-two samples can be run with the Ion 530 chip.

Qiagen's whole genome and control region NGS approaches, QIAseq Human Mitochondria panel and QIAseq Investigator Human Mitochondria Control Region panel, are built upon primers developed at AFDIL; the targets are amplified using the Qiagen Multiplex PCR Kit and can be sequenced using any Illumina NGS platform including the HiSeq, NextSeq, MiSeq, or MiniSeq. The QIAseq 1-Step Amplicon Library Kit is used to prepare DNA libraries for NGS applications. The Qiagen GeneRead Adapter I Set A 12-plex includes twelve barcoded adapters for ligation to the DNA library. The QIAseq Index Kit is used for adding indices prior to NGS. The data can be analyzed using Illumina BaseSpace apps.

The Illumina mtGenome procedure employs long PCR primers developed by Mark Wilson's lab that can be amplified using the Illumina Nextera XT library preparation kit with TaKaRa LA Taq DNA polymerase and sequenced using the Illumina MiSeq v2 (2×150 cartridge). This approach was tested on control DNA and non-probative case samples (Peck et al., 2018). Illumina's human mtDNA D-Loop hypervariable region protocol and sequencing using the MiSeq was also tested with the TaKaRa Ex Taq® HS proofreading enzyme for hair shafts (Holland, Wilson et al. 2017).

The AFDIL whole mitochondrial genome method of Fendt et al. (2009) can be performed using the Kapa Hyper Plus Library Kit to amplify the long PCR primers using TaKaRa LA Taq (GC Buffer and BSA) and sequenced using the Illumina V3 (2×300 cartridge).

7.5 Mitochondrial Sequence Data Interpretation and Reporting

Some mitochondrial chromosomes may contain mutations and differ in sequence from the other mitochondrial chromosomes in a person. Mitochondrial DNA typing can be used to screen for heteroplasmy variants in subpopulations of mitochondria (Li et al., 2010, Just et al., 2015, Holland et al., 2018). NGS has been evaluated by several groups to assess heteroplasmy and determine haplogroup (Li et al. 2010, Just et al. 2015, Holland, Pack et al. 2017, Cho et al. 2018, Kim et al. 2018, Strobl et al. 2019). Heteroplasmy is a term used to describe the occurrence of more than one type of organelle genome and can be indicative of mitochondrial diseases. In heteroplasmy, the mutation or variant mtDNA mixes with the "normal" type mtDNA in the cell. An oocyte may contain a mitochondrion with a mutation in its DNA; this mitochondrion can be replicated and the variation transmitted to new cells upon cell division. Additionally, cells can accumulate mutations over time, and the aggregate is observed in the sequencing data. The mutated mitochondrial chromosomes accumulate in the cell and can begin to predominate and influence the cellular and individual genotype. Upon fertilization of an egg cell, the heteroplasmy can be transmitted differently to different organs and tissues and produce a mosaicism. Issues with mtDNA typing including heteroplasmy need to be considered when analyzing mtDNA data. Triplasmy is a term used to describe heteroplasmy at two sites in an individual. In forensics, heteroplasmy can be used to further discriminate within a maternal line, since the variants or mutations detected in an individual may not appear in other individuals in the same maternal line. For example, in a recent study using SRM 2392 DNA, all sequence calls were concordant with the Certificate of Analysis from NIST with the exception position 64 in which heteroplasmy (ISHI poster) was detected using NGS using ForenSeq that was undetected using Sanger sequencing (Peck et al. 2018).

After sequencing, the evidence data is reported as compared to the known rCRS sequence by base position number and variant (e.g., 73A). An N is used to denote a base that cannot be unambiguously determined. Confirmed heteroplasmy is reported as R for A/G and Y for C/T. If there are insertions, a ".1" is added directly to the number position of the insertion (e.g., 315.1C for 6Cs instead of 5 following the T at 310 prior to 316). A deletion is denoted by a "D," "d," or "del" following the position where the deletion was observed (e.g., 309D, 309d, or 309del). The sequences of the questioned samples are compared to reference known samples, if available. The haplotype is determined using databases such as the European DNA profiling (EDNAP) mtDNA population database (EMPOP) and frequency is determined (Butler 2005).

Mitochondrial DNA typing data analysis of ForenSeq mtDNA Control Region and Whole Genome data can be performed with the upgraded

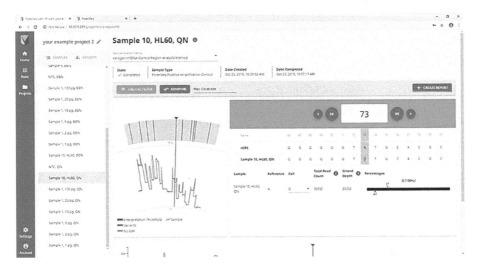

Figure 7.1 Variation between SNP 73 in HL-60 as compared to the rCRS.

Verogen Universal Analysis Software (UAS) v2. The software tool is based on mtDNA BaseSpace applications, and the mtDNA Variant Processor is an app on Illumina's BaseSpace cloud computing platform. The software analyzes the raw data retrieved from the MiSeq FGx instrument, shows a coverage map, and calls the base positions as compared to the rCRS (Figure 7.1).

The UAS can be used to assess quality metrics from a sequencing run, evaluate reads per sample, view the SNP base at each position, compare the variable positions to the rCRS, and view the number of reads at each SNP. Figure 7.2 shows the reads per sample in a graph generated by the UAS.

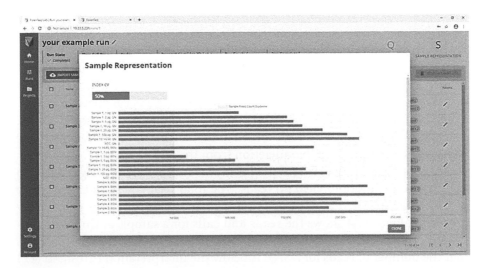

Figure 7.2 Reads per sample.

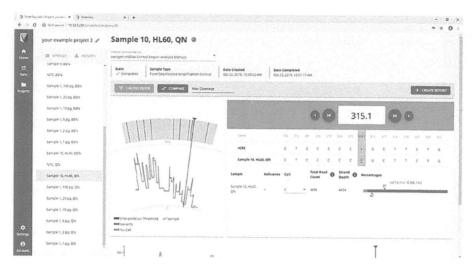

Figure 7.3 Insertion at 315.1 in the HL-60 standard as compared to the rCRS.

The mtDNA Variant Processor app performs adaptor trimming, alignment, primer, variant calling, and performs variant call format (.vcf) file output. In the trimming step, the adaptor sequences are removed from the forward and reverse reads until only three adaptor bases are found on each end of the read. Incomplete amplicons and the reads that were trimmed excessively are discarded. To perform base calling, the sequences are aligned from the true start of the circle using BWA-MEM with parameters optimized for homopolymeric C-stretches (e.g., HVII that can lead to discordance). Indels are realigned and a 3′ alignment is used in C-stretches. Figure 7.3 shows the insertion at 315.1 in the HL-60 standard as compared to the rCRS. Primer contributions are removed from the reads to accurately call the variants. Using all of the reads, the data is compiled to identify the consensus base for the allele call. The quality score is used to filter the bases prior to use in calling. A minimum read count is used for calls; locations in which the reads or noise fall below the minimum are flagged. A score is calculated for each called position based upon the BaseQ, MapQ, and analysis threshold scores. At each SNP position, the base call may match the rCRS, differ from the rCRS, or reflect an insertion or deletion from the rCRS sequence.

A full profile is expected for the positive control while the negative or no template control should yield no reads. The total read count and percentage of each base detected at each position is also numerically and graphically presented. Figure 7.4 shows the reads and calls for a sample at three concentrations (1, 5, and 100 pg) at position 489. A purple warning flag indicates that the number of reads for a SNP is below the interpretation threshold (IT)

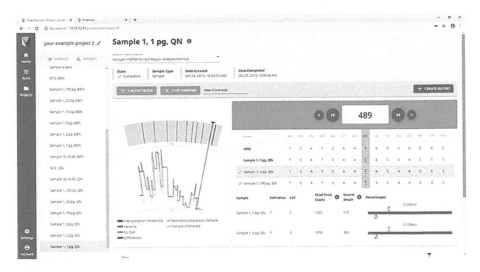

Figure 7.4 Total reads and calls for a sample at three concentrations (1, 5, and 100 pg) at position 489.

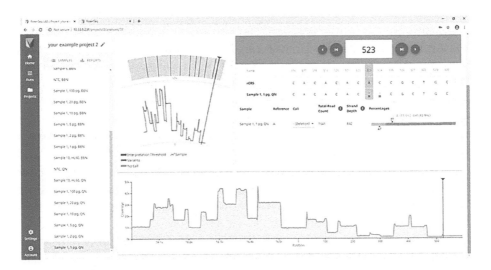

Figure 7.5 A purple warning flag indicates that the number of reads for position 523 is below the IT in a 1 pg sample.

at position 523 in the 1 pg sample (Figure 7.5). The mtDNA Variant Analyzer BaseSpace app gives a visual representation of the read coverage at each base, variations from the rCRS, modified bases, and bases flagged with colorimetric indicators and enables the user to advance the dial through the mitochondrial base locations. Filters can be used to restrict the view to the hotspots.

The Sample Compare mode opens a new window and includes the base call, paired end depth, variant quality score, and base read percentages. Any sample can be assigned as the reference and used to generate comparisons and call differences to other samples. Users can compare results within a run or those generated over time. The UAS does not have a tool for haplotypes, but the generated report can be cut and pasted into the information into EMPOP database for the prediction (https://empop.online/). Reports can be generated to summarize and output the results. The variant report can be exported to Excel. The BAM can be used for additional investigation with other tools including integrative genomics viewer (IGV) for alignments.

Mitochondrial DNA typing data analysis of Precision ID mtDNA Control Region Panel, and Whole Genome Panel can be performed in the Applied Biosystems Converge Software v2.1 NGS Data Analysis module released in 2018 which is approved for uploading the data into CODIS. Converge combines CE and NGS data analysis into one platform. Mitochondrial DNA analysis can be difficult due to mixtures, heteroplasmy, insertions, and deletions that can complicate alignments. The Precision ID mtDNA sequence data varies in read depth just as the ForenSeq sequence data does. Like the UAS, the Converge software automates base calling and alignment, detects heteroplasmy to ~10%, identifies variants, and allows labs to look at sequences throughout the control region. It can filter common variants, deleterious variants, and other variants. Converge includes annotation information from more than twenty databases. The top tabs include Home, Samples, Analyses, and Workflow. There are quick links to access samples, workflows, and analyses. By selecting the Workflow tab, the user can create a new workflow. The first step is to name the workflow and add a description. The subsequent steps include reference, annotation, filters, plugins, final report, parameters, and confirm. Data can be displayed as a variant impact tree or a copy number variation (CNV) heat map. Compared to Sanger sequencing, variant calling with SRM 2392 sequenced with NGS showed little discordance and was limited primarily to the 309 position.

Other software available for the analysis of NGS mitotyping data is the Softgenetics GeneMarker HTS, which is compatible with forensic nomenclature and can be used to produce an EMPOP formatted report. The CLC Genomics Workbench can be used to analyze AFDIL and Qiagen NGS data using the AFDIL-QIAGEN mtDNA Expert (AQME) tool. AQME generates an editable mtDNA profile consistent with forensic naming conventions and reporting information. AQME also estimates mtDNA haplogroup, has optional outputs including SNP metrics, exportable files, and is compatible with any sample type, library preparation kit, or NGS platform (Sturk-Andreaggi et al. 2017). Van Neste et al. (2014) disseminated tools including Python scripts and My-Forensic-Loci-queries (MyFLq) for mitochondrial sequence data analysis. Liu et al. (2018) performed a review of bioinformatics tools and methods for forensic DNA analysis for users seeking additional information and options.

In evaluations of the several kits and methods, variations have been observed in read depth among the methods. NIST reported on their testing of the Nextera® XT DNA Sample Preparation Kit and sequencing with the Illumina MiSeq™ in 2014 (King et al. 2014). More consistent coverage and less likelihood of dropout have been obtained using the AFDIL Whole mtGenome Method (Kapa Hyper Plus Library Kit) than the Nextera® Library kit whole genome method in a performance analysis at NIST. McElhoe and Holland (2020) reported upon the discrimination of heteroplasmy from system noise and substitution and sequence specific errors in MiSeq NGS data.

NGS has improved mixture analysis for autosomal DNA typing. Three or more heteroplasmic sites in HV1 and HV2 usually indicate a mixture, but mtDNA typing is not used for mixture analysis. In addition, nuclear pseudo-genes (chromosome 11) can amplify and contaminate mtDNA sequence; this can be reduced or eliminated by using WGA to enrich the mtDNA or DNase I digestion to eliminate nuclear DNA.

7.6 Recent Reports of Mitotyping Using NGS for Forensic Applications

There has been a flurry of interest in massively parallel sequencing (MPS) tools for mitotyping in forensic testing over the past five years. Many teams have evaluated MPS in their labs using the commercial kits we have described. Kim et al. (2015) reported on massively parallel sequencing using the 454 GS Junior instrument to sequence the mitochrondrial HVI and HVII regions, and Churchill et al. (2016) reported on the use of the Ion PGM™ for sequencing the whole mitochondrial genome and the HID-Ion STR 10-plex panel. Davis et al. (2015) reported upon the use of NGS for buccal swab, bone, and tissue samples using the Nextera® XT kit and the MiSeq and found that the profiles were concordant and that the method was able to resolve length heteroplasmy. Woerner et al. (2018) reported on the use of the Precision ID mtDNA whole genome panel using the Ion S5 and additionally tested it using the MiSeq; Strobl et al. (2018) also evaluated the Precision ID whole mtDNA genome panel but tested it on the Ion Personalized Genome Machine (PGM); Avent et al. (2019) reported on the use of the Qiagen 140-locus SNP NGS assay; and Cihlar et al. (2020) also reported on the use of the Precision ID mtDNA whole genome panel sequenced with the Ion Chef and Ion S5. Zascavage et al. (2018) evaluated the Oxford Nanopore MinION device for mtDNA analysis and found that it resulted in a 1.00% error rate in sequencing the whole mtDNA genome. Shih et al. (2018) performed probe capture instead of PCR amplification for target amplification offers for the analysis of limited and mock degraded samples and individual telogen hairs. Huszar et al. (2019) reported on their evaluation of the prototype Promega PowerSeq™ Auto/Mito/Y System which employs overlapping primers to produce short amplicons. The Promega PowerSeq™ CRM Nested System followed by Nextera® XT library preparation and NGS were used to assess heteroplasmy and DNA damage under low template conditions (Holland et al. 2021). McCord and Lee (2018) recently edited an issue of *Electrophoresis* entitled "Novel Applications of Massively Parallel Sequencing (MPS) in Forensic Analysis," which includes twenty forensic NGS reports including a several on mtDNA applications. Holland et al. (2019) reported on mtDNA profiles obtained from unfired ammunition components using MPS.

Stoljarova et al. (2016) evaluated mtDNA variation in the Estonian population using MPS. Park et al. (2017) reported on full mtDNA genome sequencing of Korean individuals using MPS, and Avila et al. (2019) reported on the same for Brazilian admixed individuals. Melchionda et al. (2020) reported on NGS DNA typing results obtained with the Romanian population and polymorphisms that have not been previously detected using other tools that enable increased haplogroup discrimination.

In November 2020, Barrio et al. reported on the CHEP-ISFG collaborative exercise using MPS. NIST reported upon implementation of mtDNA typing using NGS in that laboratory (Churchill et al. 2017). Brandhagen et al. (2020) reported on validating mtDNA typing using NGS for casework at the FBI laboratory.

7.7 Mitochondrial Sequence Data and Databases

The Federal Bureau of Investigation (FBI) hosts the Combined DNA Index System (CODIS) database for law enforcement purposes which includes mtDNA profiles. The Scientific Working Group on DNA Analysis Methods (SWGDAM) has produced new interpretation guidelines "Interpretation Guidelines for Mitochondrial DNA Analysis by Forensic DNA Testing Laboratories," which includes sections on using NGS in forensic DNA testing. The mtDNA testing interpretation document 2019 updates include sections on sequence analysis criteria, interpretation of C-stretches, and mixture interpretation. The US National DNA Index System (NDIS) Procedures Board approved (5/2/19) the use of the MiSeq FGx® Forensic Genomics System to collect data that can be entered into NDIS. Control region data collected using the Precision ID mtDNA Whole Genome Panel is also approved for inclusion in the NDIS CODIS database. Data collected using the ForenSeq and Precision ID mitotyping kits have been approved for inclusion in CODIS. A high level of accuracy is required for EMPOP submission.

Questions

1. What is mitochondrial DNA? Where is it found in the cell? Which regions are probed in forensic DNA analysis?
2. What forensic questions can mtDNA typing answer?
3. Compare and contrast Sanger sequencing and NGS methods for mtDNA DNA typing.
4. Describe the processes that are occurring in PCR1 and PCR2 in the ForenSeq mtDNA Control Region kit and compare and contrast PCR1 and PCR2 in the ForenSeq mtDNA Control Region kit and the ForenSeq DNA Signature Prep Kit.
5. Describe heteroplasmy and how this can be used in mtDNA analysis for forensic applications.
6. Explain how haplotypes are assigned using sequencing data.

References

Ambers, A., Bus, M.M., King, J.L., Jones, B., Durst, J., Bruseth, J.E., Gill-King, H., and B. Budowle. "Forensic genetic investigation of human skeletal remains recovered from the La Belle shipwreck." *Forensic Science International*, 306 (January 2020): 110050. doi:10.1016/j.forsciint.2019.110050.

Ambers, A.D., Churchill, J.D., King, J.L., Stoljarova, M., Gill-King, H., Assidi, M., Abu-Elmagd, M., Buhmeida, A., Al-Qahtani, M., and B. Budowle. "More comprehensive forensic genetic marker analyses for accurate human remains identification using massively parallel DNA sequencing." *BMC Genomics* 17, Suppl 9 (October 17, 2016): 750. doi:10.1186/s12864-016-3087-2.

Anderson, S., Bankier, A.T., Barrell, B.G., de Bruijn, M.H., Coulson, A.R., Drouin, J., Eperon, I.C., Nierlich, D.P., Roe, B.A., Sanger, F., Schreier, P.H., Smith, A.J., Staden, R., and I.G. Young. "Sequence and organization of the human mitochondrial genome." *Nature* 290, no. 9 (April 9, 1981): 457–465. doi:10.1038/290457a0.

Andrews, R.M., Kubacka, I., Chinnery, P.F., Lightowlers, R.N., Turnbull, D.M., and N. Howell. "Reanalysis and revision of the Cambridge reference sequence for human mitochondrial DNA." *Nature Genetics* 23, no. 2 (October 1999): 147. doi:10.1038/13779.

Avent, I., Kinnane, A.G., Jones, N., Petermann, I., Daniel, R., Gahan, M.E., and D. McNevin. "The QIAGEN 140-locus single-nucleotide polymorphism (SNP) panel for forensic identification using massively parallel sequencing (MPS): An evaluation and a direct-to-PCR trial." *International Journal of Legal Medicine* 133, no. 3 (May 2019): 677–688. doi:10.1007/s00414-018-1975-5.

Avila, E., Graebin, P., Chemale, G., Freitas, J., Kahmann, A., and C.S. Alho. "Full mtDNA genome sequencing of Brazilian admixed populations: A forensic-focused evaluation of a MPS application as an alternative to Sanger sequencing methods." *Forensic Science International: Genetics* 42 (2019): 154–164. doi:10.1016/j.fsigen.2019.07.004.

Barrio, P.A., García, Ó., Phillips, C., Prieto, L., Gusmão, L., Fernández, C., Casals, F., Freitas, J.M., González-Albo, M.D.C., Martín, P., Mosquera, A., Navarro-Vera, I., Paredes, M., Pérez, J.A., Pinzón, A., Rasal, R., Ruiz-Ramírez, J., Trindade, B.R., and A. Alonso. "The first GHEP-ISFG collaborative exercise on forensic applications of massively parallel sequencing." *Forensic Science International: Genetics* 49 (November 2020): 102391. doi:10.1016/j.fsigen.2020.102391.

Biesecker, L.G., Bailey-Wilson, J.E., Ballantyne, J., Baum, H.R., Bieber, F.R., Brenner, C., Budowle, B., Butler, J.M., Carmody, G., Conneally, P.M., Duceman, B., Eisenberg, A., Forman, L., Kidd, K.K. Leclair, B., Niezgoda, S., Parsons, T.J., Pugh, E., Shaler, R., Sherry, S.T., Sozer, A., and A. Walsh. "DNA identifications after the 9/11 world trade center attack." *Science* 310 no. 5751 (November 18, 2005): 1122–1123. doi:10.1126/science.1116608.

Bini, C., Ceccardi, S., Colalongo, C., Ferri, G., Falconi, M., Pelotti, S., and G. Pappalardo. "Population data of mitochondrial DNA region HVIII in 150 individuals from Bologna (Italy)." *International Congress Series* 1239 (January 2003):525–528. doi:10.1016/S0531-5131(02)00559-9.

Brandhagen, M.D., Just, R.S., and J.A. Irwin. "Validation of NGS for mitochondrial DNA casework at the FBI Laboratory." *Forensic Science International: Genetics* 44 (January 2020): 102151. doi:10.1016/j.fsigen.2019.102151.

Bruijns, B., Tiggelaar, R., and H. Gardeniers. "Massively parallel sequencing techniques for forensics: A review." *Electrophoresis* 39, no. 21 (November 2018): 2642–2654. doi:10.1002/elps.201800082.

Budowle, B., Wilson, M.R., DiZinno, J.A., Stauffer, C., Fasano, M.A., Holland, M.M., and K.L. Monson. "Mitochondrial DNA regions HVI and HVII population data." *Forensic Science International* 103, no. 1 (July 12, 1999): 25–35. doi:10.1016/S0379-0738(99)00042-0.

Budowle, B., Allard, M.W., Wilson, M.R., and R. Chakraborty. "Forensics and Mitochondrial DNA: Applications, Debates, and Foundations." *Annual Review of Genomics and Human Genetics* 4 (September 2003): 119–141. doi:10.1146/annurev.genom.4.070802.110352.

Buś, M.M., Lembring, M., Kjellström, A., Strobl, C., Zimmermann, B., Parson, W., and M. Allen. "Mitochondrial DNA analysis of a Viking age mass grave in Sweden." *Forensic Science International: Genetics* 42 (September 2019): 268–274. doi:10.1016/j.fsigen.2019.06.002

Butler, J. *Forensic DNA Typing*, 2nd ed. Burlington, MA: Elsevier Academic Press, 2005.

Chaitanya, L., Ralf, A., van Oven, M., Kupiec, T., Chang, J., Lagacé, R., and M. Kayser. "Simultaneous whole mitochondrial genome sequencing with short overlapping amplicons suitable for degraded DNA using the ion torrent personal genome machine." *Human Mutation* 36, no. 12 (December 2015): 1236–1247. doi:10.1002/humu.22905.

Chan, D.C., and E.A. Schon. "Eliminating mitochondrial DNA from sperm." *Developmental Cell* 22, no. 3 (March 13, 2012): 469–470. doi:10.1016/j.devcel.2012.02.008.

Cho, S., Kim, M. Y., Lee, J. H., and S.D. Lee. "Assessment of mitochondrial DNA heteroplasmy detected on commercial panel using MPS system with artificial mixture samples." *International Journal of Legal Medicine* 132 no. 4 (July 2018): 1049–1056. doi:10.1007/s00414-017-1755-7.

Churchill, J.D., King, J.L., Chakraborty, R., and B. Budowle. "Effects of the Ion PGM™ Hi-Q™ sequencing chemistry on sequence data quality." *International Journal of Legal Medicine* 130, no. 5 (September 2016): 1169–1180. doi:10.1007/s00414-016-1355-y.

Churchill, J.D., Peters, D., Capt, C., Strobl, C., Parson, W., and B. Budowle. "Working towards implementation of whole genome mitochondrial DNA sequencing into routine casework." *Forensic Science International: Genetics Supplement Series* 6 (December 2017): e388–e389. doi:10.1016/j.fsigss.2017.09.167.

Cihlar, J.C., Amory, C., Lagacé, R., Roth, C., Parson, W., and B. Budowle. "Developmental validation of a MPS workflow with a PCR-based short amplicon whole mitochondrial genome panel." *Genes (Basel)* 11, no. 11 (November 13, 2020): 1345. doi:10.3390/genes11111345.

Coble, M.D., Just, R.S., O'Callaghan, J.E., Letmanyi, I.H., Peterson, C.T., Irwin, J.A., and T.J. Parons. "Single nucleotide polymorphisms over the entire mtDNA genome that increase the power of forensic testing in Caucasians." *International Journal of Legal Medicine* 118 (February 4, 2004): 137–146. doi:10.1007/s00414-004-0427-6.

Cuenca, D., Battaglia, J., Halsing, M, and S. Sheehan. "Mitochondrial Sequencing of Missing Persons DNA casework by implementing thermo fisher's Precision ID mtDNA whole genome assay." *Genes (Basel)* 4, no. 11 (November 2020): 1303. doi:10.3390/genes11111303.

Davis, C., Peters, D., Warshauer, D., King, J., and B. Budowle. "Sequencing the hypervariable regions of human mitochondrial DNA using massively parallel sequencing: Enhanced data acquisition for DNA samples encountered in forensic testing." *Legal Medicine (Tokyo, Japan)* 17, no. 2 (March 2015): 123–127. doi:10.1016/j.legalmed.2014.10.004.

Dobrowolski, S.F., Gray, J., Miller, T., and M. Sears. "Identifying sequence variants in the human mitochondrial genome using high-resolution melt (HRM) profiling." *Human Mutation* 30, no. 6 (June 2009): 891–898. doi:10.1002/humu.21003.

Eduardoff, M., Xavier, C., Strobl, C., Casas-Vargas, A., and W. Parson. "Optimized mtDNA control region primer extension capture analysis for forensically relevant samples and highly compromised mtDNA of different age and origin." *Genes (Basel)* 8, no. 10 (September 21, 2017): 237. doi:10.3390/genes8100237.

Elkins, K.M. *Forensic DNA Biology: A Laboratory Manual*. Waltham, MA: Elsevier Academic Press, 2013.

Fendt, L., Zimmerman, B., Daniaux, M., and W. Parson. "Sequencing strategy for the whole mitochondrial genome resulting in high quality sequences." *BMC Genomics* 10 (March 30, 2009): 139. doi:10.1186/1471-2164-10-139.

ForenSeq mtDNA Control Region Kit Reference Guide, Document # VD 2019001 Rev. A. Accessed May 22, 2021. https://verogen.com/wp-content/uploads/2019/08/ForenSeq-mtDNA-Control-Region-Guide-VD2019001-A.pdf

Fridman, C., and R.S. Gonzalez. "HVIII discrimination power to distinguish HVI and HVII common sequences." *Forensic Science International: Genetics Supplement Series* 2 (December 2009): 320–321. doi:10.1016/j.fsigss.2009.07.011.

Gaag, K.J.V., Desmyter, S., Smit, S., Prieto, L., and T. Sijen. "Reducing the number of mismatches between Hairs and Buccal references when analysing mtDNA heteroplasmic variation by massively parallel sequencing." *Genes (Basel)* 11, no. 11 (November 16, 2020): 1355. doi:10.3390/genes11111355.

Gabriel, M.N., Huffine, E.F., Ryan, J.H., Holland, M.M., and T.J. Parsons. "Improved MtDNA sequence analysis of forensic remains using a 'mini-primer set' amplification strategy." *Journal of Forensic Sciences* 46, no. 2 (March 2001): 247–253.

Gallimore, J.M., McElhoe, J.A., and M.M. Holland. "Assessing heteroplasmic variant drift in the mtDNA control region of human hairs using an MPS approach." *Forensic Science International: Genetics* 32 (January 2018): 7–17.

Hickman, M.P., Grisedale, K.S., Bintz, B.J., Burnside, E.S., Hanson, E.K., Ballantyne, J., and M.R. Wilson. "Recovery of whole mitochondrial genome from compromised samples via multiplex PCR and massively parallel sequencing." *Future Science OA* 4, no. 9 (August 24, 2018): FSO336. doi:10.4155/fsoa-2018-0059.

Holland, M.M., and T.J. Parsons. "Mitochondrial DNA sequence analysis – Validation and use for forensic casework." *Forensic Science Reviews* 11, no. 1 (June 1999): 21–50.

Holland, M.M., Pack, E.D., and J.A. McElhoe. "Evaluation of GeneMarker® HTS for improved alignment of mtDNA MPS data, haplotype determination, and heteroplasmy assessment." *Forensic Science International: Genetics* 28 (May 2017): 90–98. doi:10.1016/j.fsigen.2017.01.016.

Holland, M.M., Wilson, L.A., Copeland, S., Dimick, G., Holland, C.A., Bever, R., and J.A. McElhoe. "MPS analysis of the mtDNA hypervariable regions on the MiSeq with improved enrichment." *International Journal of Legal Medicine* 131, 4 (July 2017): 919–931. doi:10.1007/s00414-017-1530-9.

Holland, M.M., Makova, K.D., and J.A. McElhoe. "Deep-coverage MPS analysis of heteroplasmic variants within the mtGenome allows for frequent differentiation of maternal relatives." *Genes* 9, no. 3 (February 26, 2018): 124. doi:10.3390/genes9030124.

Holland, M.M., Bonds, R.M., Holland, C.A., and J.A. McElhoe. "Recovery of mtDNA from unfired metallic ammunition components with an assessment of sequence profile quality and DNA damage through MPS analysis." *Forensic Science International: Genetics* 39 (March 2019): 86–96. doi:10.1016/j.fsigen.2018.12.008.

Holland, C.A., McElhoe, J.A., Gaston-Sanchez, S., and M.M. Holland. "Damage patterns observed in mtDNA control region MPS data for a range of template concentrations and when using different amplification approaches." *International Journal of Legal Medicine* 135, no. 1 (January 2021): 91–106. doi:10.1007/s00414-020-02410-0.

Huszar, T.I., Wetton, J.H., and M.A. Jobling. "Mitigating the effects of reference sequence bias in single-multiplex massively parallel sequencing of the mitochondrial DNA control region." *Forensic Science International: Genetics* 40 (May 2019): 9–17. doi:10.1016/j.fsigen.2019.01.008.

Ivanov, P.L., Wadhams, M.J., Roby, R.K., Holland, M.M., Weedn, V.W., and T.J. Parsons. "Mitochondrial DNA sequence heteroplasmy in the Grand Duke of Russia Georgij Romanov establishes the authenticity of Tsar Nicholas II." Nature Genetics 12, no. 4 (April 1996): 417–420. doi:10.1038/ng0496-417.

Just, R.S., Irwin, J.A., and W. Parson. "Mitochondrial DNA heteroplasmy in the emerging field of massively parallel sequencing." *Forensic Science International: Genetics* 18 (September 2015): 131–139. doi:10.1016/j.fsigen.2015.05.003.

Kim, H., Erlich, H.A., and C.D. Calloway. "Analysis of mixtures using next generation sequencing of mitochondrial DNA hypervariable regions." *Croatian Medical Journal* 56, no. 3 (May 31, 2015): 208–217. doi:10.3325/cmj.2015.56.208.

Kim, M.Y., Cho, S., Lee, J.H., Seo, H.J., and S.D. Lee. "Detection of innate and artificial mitochondrial DNA heteroplasmy by massively parallel sequencing: Considerations for analysis." *Journal of Korean Medical Science* 33, no. 52 (December 11, 2018): e337. doi:10.3346/jkms.2018.33.e337.

Kim, B.M., Hong, S.R., Chun, H., Kim, S., and K.J. Shin. "Comparison of whole mitochondrial genome variants between hair shafts and reference samples using massively parallel sequencing." *International Journal of Legal Medicine* 134, no. 3 (May 2020): 853–861. doi:10.1007/s00414-019-02205-y.

King, J.L., LaRue, B.L., Novroski, N.M., Stoljarova, M., Seo, S.B., Zeng, X., Warshauer, D.H., Davis, C.P., Parson, W., Sajantila, A., and B. Budowle. "High-quality and high-throughput massively parallel sequencing of the human mitochondrial genome using the Illumina MiSeq." *Forensic Science International: Genetics* 12 (September 2014): 128–135. doi:10.1016/j.fsigen.2014.06.001.

LaBerge, G.S., Shelton, R.J., Danielson, P.B. "Forensic utility of mitochondrial DNA analysis based on denaturing high-performance liquid chromatography." *Croatian Medical Journal* 44, no. 3 (2003): 281–88.

Lee, E.Y., Lee, H.Y., Oh, S.Y., Jung, S.-E., Yang, I.S., Lee, Y.-H., Yang, W.I., and K.-J. Shin. "Massively parallel sequencing of the entire control region and targeted coding region SNPs of degraded mtDNA using a simplified library preparation method." Department of Forensic Medicine; Yonsei University College of Medicine. 2015. http://forensic.yonsei.ac.kr/presentation/116.pdf.

Levin, B.C., Hancock, D.K., Holland, K.A., Cheng, H., and K.L. Richie. "Human mitochondrial DNA—amplification and sequencing standard reference materials—SRM 2392 and SRM 2392-I." *NIST Special Publication* 260-155 (2003): 1–93.

Li, M., Schönberg, A., Schaefer, M., Nasidze, I., and M. Stoneking. "Detecting heteroplasmy from high-throughput sequencing of complete human mitochondrial DNA genomes." *American Journal of Human Genetics* 87, no. 2 (August 13, 2010): 237–249. doi:10.1016/j.ajhg.2010.07.014.

Liu, Y.Y., and S. Harbison. "A review of bioinformatic methods for forensic DNA analyses." *Forensic Science International: Genetics* 33 (March 2018): 117–128. doi:10.1016/j.fsigen.2017.12.005.

Ma, K., Zhao, X., Li, H., Cao, Y., Li, W., Ouyang, J., Xie, L., and W. Liu. "Massive parallel sequencing of mitochondrial DNA genomes from mother-child pairs using the ion torrent personal genome machine (PGM)." *Forensic Science International: Genetics* 32 (January 2018): 88–93. doi:10.1016/j.fsigen.2017.11.001.

Marshall, C., Sturk-Andreaggi, K., Daniels-Higginbotham, J., Oliver, R.S., Barritt-Ross, S., and T.P McMahon. "Performance evaluation of a mitogenome capture and Illumina sequencing protocol using non-probative, case-type skeletal samples: Implications for the use of a positive control in a next-generation sequencing procedure." *Forensic Science International: Genetics* 31 (November 2017): 198–206. doi:10.1016/j.fsigen.2017.09.001.

McCord, B., and S.B. Lee. "Novel applications of Massively Parallel Sequencing (MPS) in forensic analysis." *Electrophoresis* 39, no. 21 (November 2018): 2639–2641. doi:10.1002/elps.201870175.

McElhoe, J.A., Holland, M.M., Makova, K.D. Su, M.S., Paul, I.M., Baker, C.H., Faith, S.A., and B. Young. "Development and assessment of an optimized next-generation DNA sequencing approach for the mtgenome using the Illumina MiSeq." *Forensic Science International: Genetics* 13 (November 2014) 20–29. doi:10.1016/j.fsigen.2014.05.007.

McElhoe, J.A., and M.M. Holland. "Characterization of background noise in MiSeq MPS data when sequencing human mitochondrial DNA from various sample sources and library preparation methods." *Mitochondrion* 52 (May 2020): 40–55. doi:10.1016/j.mito.2020.02.005.

Melchionda, F., Stanciu, F., Buscemi, L., Pesaresi, M., Tagliabracci, A., and C. Turchi. "Searching the undetected mtDNA variants in forensic MPS data." *Forensic Science International: Genetics* 49 (November 2020):102399. doi:10.1016/j.fsigen.2020.102399.

MITOMAP: mtDNA Control Region Sequence Variants. Accessed January 25, 2021. https://www.mitomap.org/foswiki/bin/view/MITOMAP/Polymorphisms Control.

Park, S., Cho, S., Seo, H.J., Lee, J.H., Kim, M.Y., and S.D. Lee. "Entire mitochondrial DNA sequencing on massively parallel sequencing for the Korean population." *Journal of Korean Medical Science* 32, no. 4 (April 2017): 587–592. doi:10.3346/jkms.2017.32.4.587.

Parson, W., Huber, G., Moreno, L., Madel, M.B., Brandhagen, M.D., Nagl, S., Xavier, C., Eduardoff, M., Callaghan, T.C., and J.A. Irwin. "Massively parallel sequencing of complete mitochondrial genomes from hair shaft samples." *Forensic Science International: Genetics* 15 (March 2015): 8–15. doi:10.1016/j.fsigen.2014.11.009.

Peck, M.A., Sturk-Andreaggi, K., Thomas, J. T., Oliver, R.S., Barritt-Ross, S., and C. Marshall. "Developmental validation of a Nextera XT mitogenome Illumina MiSeq sequencing method for high-quality samples." *Forensic Science International: Genetics* 34 (May 2018): 25–36. doi:10.1016/j.fsigen.2018.01.004.

Quintáns, B., Álvarez-Iglesias, V., Salas, A., Phillips, C., Lareu, M.V., and A. Carracedo. "Typing of mitochondrial DNA coding region SNPs of forensic and anthropological interest using SNaPshot minisequencing." *Forensic Science International* 140, no. 2–3 (March 10, 2004): 251–257. doi:10.1016/j.forsciint.2003.12.005.

Rathbun, M.M., McElhoe, J.A, Parson, W., and M.M. Holland. "Considering DNA damage when interpreting mtDNA heteroplasmy in deep sequencing data." *Forensic Science International: Genetics* 26 (January 2017): 1–11. doi:10.1016/j.fsigen.2016.09.008.

Remualdo, V.R., and R.N. Oliveira. "Analysis of mitochondrial DNA from the teeth of a cadaver maintained in formaldehyde." *The American Journal of Forensic Medicine and Pathology* 28, no. 2 (June 2007): 145–146. doi:10.1097/PAF.0b013e31805f67d1.

Seo, Y., Uchiyama, T., Matsuda, H., Shimizu, K., Takami, Y., Nakayama, T., and K. Takahama. "Mitochondrial DNA and STR typing of matter adhering to an earphone." *Journal of Forensic Sciences* 47, no. 3 (May 2002): 605–608.

Shih, S.Y., Bose, N., Gonçalves, A.B.R., Erlich, H.A., and C.D, Calloway. "Applications of probe capture enrichment next generation sequencing for whole mitochondrial genome and 426 nuclear SNPs for forensically challenging samples." *Genes (Basel)* 9, no. 1 (January 22, 2018): 49. doi:10.3390/genes9010049.

Strobl, C., Eduardoff, M., Bus, M.M., Allen, M., and W. Parson. "Evaluation of the precision ID whole MtDNA genome panel for forensic analyses." *Forensic Science International: Genetics* 35 (July 2018): 21–25. doi:10.1016/j.fsigen.2018.03.013.

Strobl, C., Churchill Cihlar, J., Lagacé, R., Wootton, S., Roth, C., Huber, N., Schnaller, L., Zimmermann, B., Huber, G., Lay Hong, S., Moura-Neto, R., Silva, R., Alshamali, F., Souto, L., Anslinger, K., Egyed, B., Jankova-Ajanovska, R., Casas-Vargas, A., Usaquén, W., Silva, D., Barletta-Carrillo, C., Tineo, D.H., Vullo, C., Würzner, R., Xavier, C., Gusmão, L., Niederstätter, H., Bodner, M., Budowle, B., and W. Parson. "Evaluation of mitogenome sequence concordance, heteroplasmy detection, and haplogrouping in a worldwide lineage study using the Precision ID mtDNA Whole Genome Panel." *Forensic Science International: Genetics* 42 (September 2019): 244–251. doi:10.1016/j.fsigen.2019.07.013.

Stoljarova, M., King, J.L., Takahashi, M., Aaspõllu, A., and B. Budowle. "Whole mitochondrial genome genetic diversity in an Estonian population sample." *International Journal of Legal Medicine* 130, no. 1 (January 2016): 67–71. doi:10.1007/s00414-015-1249-4.

Sturk-Andreaggi, K., Peck, M.A., Boysen, C., Dekker, P., McMahon, T.P., and C.K. Marshall. "AQME: A forensic mitochondrial DNA analysis tool for next-generation sequencing data." *Forensic Science International: Genetics* 31 (November 2017): 189–197. doi:10.1016/j.fsigen.2017.09.010.

Vallone, P.M., Just, R.S., Coble, M.D., Butler, J.M., and T.J. Parsons. "A multiplex allele-specific primer extension assay for forensically informative SNPs distributed throughout the mitochondrial genome." *International Journal of Legal Medicine* 118, no. 3 (June 2004): 147–157. doi:10.1007/s00414-004-0428-5.

Van Neste, C., Vandewoestyne, M., Van Criekinge, W., Deforce, D., and F. Van Nieuwerburgh. "My-Forensic-Loci-queries (MyFLq) framework for analysis of forensic STR data generated by massive parallel sequencing." *Forensic Science International: Genetics* 9 (March 2014): 1–8. doi:10.1016/j.fsigen.2013.10.012.

Walichiewicz, P., Eagles, J., Daulo, A., Didier, M., Edwards, C., Fleming, K., Han, Y., Hill, T., Li, S., Rensfield, A., Sa, D., Husbands, J., Holt, C., and K. Stephens. "Performance evaluation of the ForenSeq mtDNA control region solution." ISHI, 2019. https://verogen.com/wp-content/uploads/2019/11/Mito-ISHI-Poster_final.pdf.

Wallace, D.C., Brown, M.D., and M.T. Lott. "Mitochondrial DNA variation in human evolution and disease." *Gene* 238 (September 30, 1999): 211–230. doi:10.1016/S0378-1119(99)00295-4.

Wang, Z., Zhang, S., Bian, Y., and C. Li. "Differentiating between monozygotic twins in forensics through next generation mtGenome sequencing." *Forensic Science International: Genetics Supplement Series* 5 (September 10, 2015): e58–e59. doi:10.1016/j.fsigss.2015.09.023.

Wang, Z., Zhu, R., Zhang, S., Bian, Y., Lu, D., and C. Li. "Differentiating between monozygotic twins through next-generation mitochondrial genome sequencing." *Analytical Biochemistry* 490 (December 1, 2015): 1–6. doi:10.1016/j.ab.2015.08.024.

Warner, J.B., Bruin, E.J., Hannig, H., Hellenkamp, F., Hörning, A., Mittmann, K., van der Steege, G., de Leij, L.F.M.H., and H.S.P. Garritsen. "Use of sequence variation in three highly variable regions of the mitochondrial DNA for the discrimination of allogeneic platelets." *Transfusion* 46 (2006): 554–561. doi:10.1111/j.1537-2995.2006.00775.x.

Woerner, A.E., Ambers, A., Wendt, F.R., King, J.L., Moura-Neto, R.S., Silva, R., and B. Budowle. "Evaluation of the precision ID mtDNA whole genome panel on two massively parallel sequencing systems." *Forensic Science International: Genetics* 36 (September 2018): 213–224. doi:10.1016/j.fsigen.2018.07.015.

Zascavage, R.R., Thorson, K., and J.V. Planz. "Nanopore sequencing: An enrichment-free alternative to mitochondrial DNA sequencing." *Electrophoresis* 40, no. 2 (January 2019): 272–280. doi:10.1002/elps.201800083.

Zhang, C., Li, H., Zhao, X., Ma, K., Nie, Y., Liu, W., Jiao, H., and H. Zhou. "Validation of expressmarker mtDNA-SNP60: A mitochondrial SNP kit for forensic application." *Electrophoresis* 37 (2018): 2848–2861. doi:10.1002/elps.201600042.

Microbial Applications of Next Generation Sequencing for Forensic Investigations

8.1 Introduction to Microbial DNA Profiling

In addition to human DNA typing using genomic DNA and mitochondrial DNA, next generation sequencing (NGS) can be used for many additional applications relevant to forensic investigations, many of which have been demonstrated in the past few years while others continue to be introduced. Research from the ten-year Human Microbiome Project (HMP) has demonstrated that microbe populations from different sources vary and can be used to attribute the community to the source (Belizário and Napolitano 2015). Similarly, microbe communities have been shown to be impacted by human disease and infection characteristics that can be differentiating (Gurenlian 2007, Chen et al. 2014, Kistler et al. 2015, Lipowski et al. 2017). Emerging NGS applications include microbial DNA profiling of a variety of forensically relevant samples including soil, blood, hair, skin, oral, nasal, vaginal, and anal sources which can aid investigators in determining the circumstances of a case (Tridico et al. 2014, Belizário and Napolitano 2015, Giampaoli et al. 2017, Leong et al. 2017, Quaak et al. 2018, Schmedes et al. 2018, Rajan et al. 2019, Woerner et al. 2019). Analysis of the human microbiome can provide scientists with an additional measure of individual identification based, at least in part, on lifestyle and behavioral patterns. The human microbiome consists of all of the microbiological organisms on the skin and within the body in the colon, vaginal cavity, mouth, and ears. Microbiome analysis can complement traditional serological analysis and be used to determine which body site a sample came from. Microbiome profiling can aid in determining the circumstances of a case including who touched a computer keyboard, which regions came into contact during sexual intercourse, and to which body region a fluid or stain came into contact with. Microbiological analysis can also be used to estimate the postmortem interval (PMI). This chapter aims to trace progress in applying NGS for microbial community profiling with a focus on forensic applications to serve as a resource of publications published to date in this area for forensic researchers and practitioners.

DOI: 10.4324/9781003196464-8   117

8.2 Why NGS?

Whereas bacteria have long been identified through culture and microscopy, biochemical enzymatic assays, and spectroscopic methods and more recently by mass spectrometry (Elkins 2019, Franco-Duarte et al. 2019, Elkins and Bender 2020, Bender et al. 2020), molecular diagnostic methods using PCR are more sensitive (Welinder-Olsson et al. 2007). PCR-based molecular fingerprinting methods and NGS can be used to understand the structure and function relationships in microbial communities in human-associated systems (Phadke et al. 2017). NGS can capture subtle differences between bacterial communities in samples without reliance on target genetic marker systems (Sjödin et al. 2012).

8.3 The Human Microbiome Project

Humans are host to microorganisms including pathogenic and non-pathogenic bacterial species that have coevolved with the immune system and aid in functions including breakdown of biomolecules in the intestines (Belizário and Napolitano 2015). The US National Institutes of Health (NIH) sponsored a decade-long study, the "Human Microbiome Project," from 2007 to 2016 with the goal of understanding how the microbiome affects human health and disease by characterizing the abundance, diversity, and functionality of microbes (Belizário and Napolitano 2015). Samples from volunteers were collected from the mouth, tonsils, placenta at birth, vagina, skin, gut, and stool (Belizário and Napolitano 2015). The bacterial families and species present in the samples were characterized, and the differentiating families of bacteria present in the samples were documented (Belizário and Napolitano 2015). The study showed that human gut bacteria were found to express over 3.3 million bacterial genes compared to the 20,000 genes expressed by the human genome or 165 times as many bacterial genes as human genes (Belizário and Napolitano 2015).

8.4 Sampling and Processing

The overall NGS approach is very similar to those applied to forensic DNA typing of human autosomal, sex-linked, and mitochondrial DNA polymorphisms. Samples must be collected using sterile and DNA-free swabs or collection devices and handled so that DNA from the investigator is not transferred to the material. The sampling device should be labeled, catalogued, and stored under cool, dry conditions until processing begins. A DNA extraction method is used to recover the cells from the microbial community on the swab/device or from the soil and extract the DNA from the cells. The quantity and quality of the extracted DNA is determined, and the extract is concentrated or diluted, as needed, for input in library preparation. More detail on DNA extraction and quantitation is found in Chapter 3. NGS instruments, including the Roche 454, Illumina HiSeq and MiSeq, and ThermoFisher Ion Proton, can be used for sequencing (Budowle et al. 2014, Clarke et al. 2017, Minogue et al. 2019). Details regarding the chemistries and instruments can be found in Chapter 2.

In a process study, buccal, nasal, and ear swabs were used to evaluate the Promega DNA IQ™ Casework Pro Kit with the Maxwell® 16 and GenElute Bacterial Genomic DNA kit to extract DNA from ten bacterial species followed by sequencing using the Ion S5 NGS system and data analysis using the metagenomics workflow in the Ion Reporter Software (Alessandrini et al. 2019). The authors were able to simultaneously purify and identify both microbial and human DNA (Alessandrini et al. 2019). This is critical in many forensic cases in which the sample quantity is often trace or limited (Alessandrini et al. 2019). In the study, the DNA IQ™ Casework Pro Kit with the Maxwell® 16 performed better than the GenElute method (Alessandrini et al. 2019).

8.5 NGS Methodology in Microbial Forensics

While single source or human mixture samples have been routinely analyzed for many years, identifying the bacterial species in a human sample is a newer investigative tool. In bacterial studies, amplifying and sequencing the 16S rRNA gene is a standard approach for microbial profiling. Alternatively whole genome shotgun (WGS) metagenomics can be used. The 16S rRNA gene is highly species specific (Franco-Duarte et al. 2019). The principle of 16S rRNA testing relies on detecting differences in the reverse-transcribed highly variable region 16S rRNA sequences (16S rDNA) to construct the microbial flora composition. Not only is sequencing faster than culture in many cases, it is also preferred in cases in which the sample is unculturable (Gilchrist et al. 2015, Franco-Duarte et al. 2019, Willis and Gabaldón 2020). NGS can be used for detecting and identifying microorganisms in research and clinical diagnosis using approaches including DNA barcoding, single cell sequencing with whole genome amplification (WGA), whole metagenome shotgun (WMS) sequencing, meta-transcriptomics, and metagenomics (Franco-Duarte et al. 2019, Willis and Gabaldón 2020). The researcher can investigate the composition of a sample including the major and minor bacterial constituents with nearly limitless multiplex ability (Minogue et al. 2019).

The 16S rRNA gene primers are designed to target the 16S rRNA genes and tagged for adding barcodes and adaptor sequences in a subsequent PCR step. The 16S rRNA contains nine hypervariable regions spanning 1500 bp flanked by conserved sequences. It is part of the 30S subunit of the prokaryotic ribosome. Because it is essential for protein synthesis, it is widely conserved among bacteria and archaea. The primer sequences are purchased, reconstituted in nuclease-free water, quantified, and diluted to the appropriate stock concentration. A master mix containing DNA polymerase and other PCR reagents is combined with the primers and the input DNA to amplify and tag the target sequences. In a second PCR step, the barcodes and adaptor sequences are added. The libraries are cleaned-up, normalized, and pooled for sequencing. The libraries are added to the flow cell, chip, or system chosen for sequencing. In human microbiome analysis, whole populations of microorganisms can be analyzed in a sample using NGS via its deep sequencing capability.

8.6 Results from the Human Microbiome Project

In the HMP, samples from the oral mucosa, skin, vaginal mucosa, and gut were analyzed by sequencing the 16S rRNA gene using NGS and identifying the metabolic genes and proteins using computational methods (Pasolli et al. 2019). Altogether the samples included ~150,000 microbial genomes attributed to 4930 species (Pasolli et al. 2019). The data was used to develop genome scale metabolic reconstructions and constraint-based modeling methods led to phenotype prediction, microbe-microbe, microbe-host, and microbe-diet metabolic interactions, and disease etiology data (Chowdhury and Fong 2020). The data demonstrated that the four sites varied in bacterial colonizers (Chowdhury and Fong 2020).

Analysis of infant body sites also showed variation in the relative abundances of key phyla in the oral mucosa, nares, gastrointestinal tract, and skin (Milani et al. 2017). The human microbiomes are first established in the infant gut and skin during pregnancy and delivery. After delivery, the microbiome changes with the new environment and the proportions of the bacterial types shift. Factors that influence gut composition in infants and children include the mode of delivery, gestational age, mode of feeding (breastmilk or formula), family members, host interactions, maternal diet, and geographical location. The most common types of bacteria encountered in their study were actinobacteria, bacteroidetes, firmicutes, fusobacteria, and proteobacteria. The most common bacteria varied for the body regions sampled. Bacteroidetes were most common in the gut microbiome. Actinobacteria were most common in the skin microbiome. Proteobacteria were the most common in the placenta microbiome, and firmicutes were most common in the vaginal microbiome. The oral microbiome contained relatively equal proportions of all of the types with a smaller proportion of actinobacteria. Saliva bacteria from HMP samples were quantified and identified using the GENIUS tool (Hasan et al. 2014). Analysis of the urine microbiome led to the determination that urine contains bacteria and viruses and is not sterile (Wojciuk et al. 2019). These sites and human body fluid samples are routinely profiled in forensic investigations for human identification applications. The HMP data and approaches demonstrate additional opportunities for sample and human individualization using NGS but analysts must be aware of the potential for bias in sequencing methods as demonstrated in Figure 8.1. A normalized comparison of the samples sequenced with the three methods is shown in Figure 8.2 in tree mode.

8.7 HMP Applications for Forensic Science

Applications of human individualization using microbiome analysis in forensic cases include typing the intestinal microbiota in infants, body fluids including oral, vaginal and fecal samples, and skin and hair samples. The intestinal microbiota was investigated using SOLiD 16S rRNA gene sequencing and SOLiD shotgun sequencing and compared to Sanger paired end sequencing reads and 454 sequencing reads with comparable results but the SOLiD sequencing yielding better resolution to the species level (Mitra et al. 2013). Variations in the infant intestinal microbiota through 16S rRNA gene fecal testing were assayed in sudden infant death syndrome (SIDS) cases as compared to healthy, age-matched controls; no significant difference was found (Leong et al. 2017). The identification of vaginal fluid is very important in forensic investigations and sexual assault cases. Human microbiome analysis was demonstrated to identify vaginal fluid (Giampaoli et al. 2017). In a recent study, DNA was extracted from eighteen samples including vaginal, oral, fecal, and yogurt and analyzed using NGS and a traditional PCR-based method; the NGS method was shown to detect more species and lead to more probative data (Giampaoli et al. 2017).

Teeth can develop biofilms formed when oral bacteria form a sticky layer on a tooth's surface (Gurenlian 2007). If not frequently removed, the biofilm can lead to the formation of dental caries, gingivitis, and periodontitis (Gurenlian 2007). Periodontitis is a common oral bacterial disease in humans; left untreated it can lead to tooth loss (Kistler et al. 2015). To better study biofilm formation on teeth, a model was seeded with natural saliva from volunteers and extracted microbial DNA 16S rRNA targets were sequenced using the 454 and analyzed using mothur (Kistler et al. 2015). Biofilm grown from the same panel of volunteers were found to be highly similar and clustered in PCA plots with the differences between panels being more pronounced after

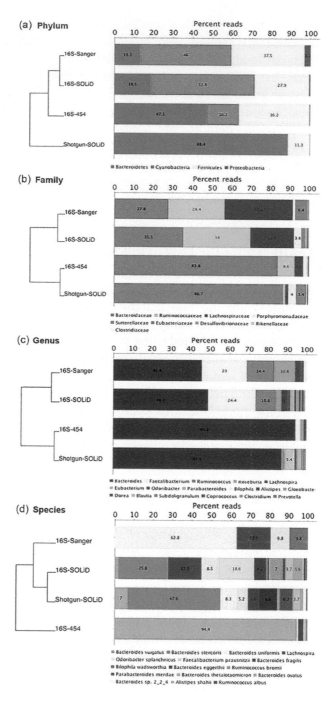

Figure 8.1 Bias in sequencing in the gut microbiome using Sanger, 454, SOLiD and Shotgun-SOLiD sequencing methods. (Suparna Mitra, Karin Förster-Fromme, Antje Damms-Machado, Tim Scheurenbrand, Saskia Biskup, Daniel H. Huson, Stephan C. Bischoff (CC 2.0). https://pubmed.ncbi.nlm.nih.gov/24564472/.)

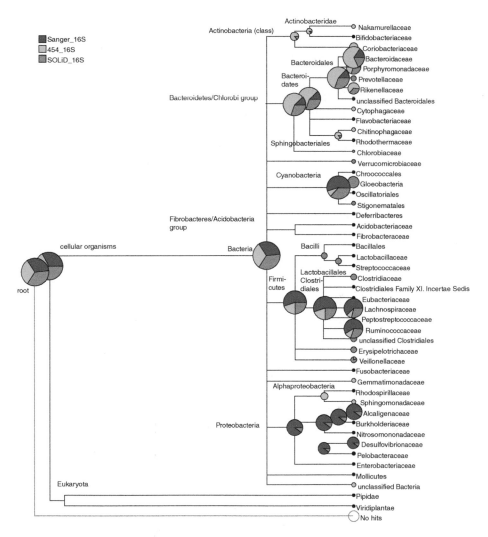

Figure 8.2 Normalized comparison between 16S samples obtained using three technologies: "'Sanger,'" "'16S-454,'" and "16S-SOLiD" datasets. Normalized comparison result obtained using MEGAN for "Sanger"-dataset (blue), "16S-454" dataset (cyan), and "16S-SOLiD" dataset (magenta) without considering "No hits" node. The tree is collapsed at "family" level of NCBI taxonomy. Circles are scaled logarithmically to indicate the number of summarized reads. (Suparna Mitra, Karin Förster-Fromme, Antje Damms-Machado, Tim Scheurenbrand, Saskia Biskup, Daniel H. Huson, Stephan C. Bischoff (CC 2.0). https://pubmed.ncbi.nlm.nih.gov/24564472/#&gid=article-figures&pid=figure-3-uid-2.)

14 days as compared to seven days (Kistler et al. 2015). A supervised learning computational method was developed and applied to predict periodontitis phenotypes based on microbial composition determined using 16S rRNA sequence data (Gurenlian 2007). The reported jackknife accuracy was 94.83% demonstrating its strength in predicting the disease status (Gurenlian 2007). In another study, pyrosequencing was used to analyze the oral microbiome and predict biofilm infections (Siqueira et al. 2012). Chen et al. (2014) suggested the approach could be applied to forensics and other research questions.

Phylogenetic profiles of microorganisms from fecal samples depended upon sequencing depth and NGS analysis method underscoring the need to develop SOPs, reference databases, and standardized bioinformatics approaches (Rajan et al. 2019). Sequencing depth beyond 60 million reads using the Illumina HiSeq 2500 was not found to improve classification (Rajan et al. 2019). The multiplex hidSkinPlex was developed for forensic identification and prediction of body site using the skin microbiome (Schmedes et al. 2018, Woerner et al. 2019). Prediction of body site using the skin microbiome had an accuracy of up to 86% (Schmedes et al. 2018). Tests were conducted using hand, chest, foot, and "all" sites (Woerner et al. 2019). A recent review and meta-analysis of research on using the skin microbiome as a forensic tool cautions that while skin microbial communities are personalized, body sites and sample time impact the profile, and although intrapersonal differences are smaller than interpersonal ones, understanding the variability will be essential to the use of this tool (Tozzo et al. 2020). NGS can be used to determine tissue type (Aly et al. 2015). The cell type from different body sites including hand, foot, groin, penis, vagina, mouth, and feces were analyzed (Quaak et al. 2018). Oral and fecal sites were clearly distinguished from skin and vaginal samples, and human feces were differentiable from dog and cat animal feces (Quaak et al. 2018). However, differentiating samples from some skin sites was difficult as some penis site samples were highly similar to vaginal samples and skin samples (Quaak et al. 2018). In another study, NGS metagenomic analysis was applied to bacteria on human scalp and pubic hair using the Roche GS Junior™ (Tridico et al. 2014). As human hairs without a root are often problematic for identification as it is difficult to obtain STR profiles, microbial NGS offers a new opportunity for discrimination (Tridico et al. 2014). The microbiomes from male and female pubic hairs were found in separate clusters using principle component analysis (PCA) (Tridico et al. 2014). The microbiome was similar for a cohabitating couple who engaged in sexual intercourse (Tridico et al. 2014). The study concluded that metagenomics was most promising for pubic hair analyses to augment STR DNA typing and mtDNA sequencing (Tridico et al. 2014). NGS can also sequence minute and degraded samples and enable better mixture analysis (Aly et al. 2015).

8.8 NGS Applications in Geolocation, Autopsy, PMI, and Lifestyle Analysis

In addition to body fluid and site identification, human microbiome analysis can be used to answer several other forensic questions. Metagenomics can answer questions about past events (Giampaoli et al. 2018). Human microbiome analysis can be used for human identification and geolocation including location of clandestine graves as well as indoor and outdoor sites (Clarke et al. 2017, Alessandrini et al. 2019). NGS can analyze forensically relevant environmental samples such as soil and water (Budowle et al. 2014, Gilchrist et al. 2015). A study of the 18S rRNA from fungi of eleven samples from different soil environments with different flora including forests, fields, grasslands, and urban park yielded nine GI matches unique to the sampling area using analysis from mothur and Blastn (Lilje et al. 2013).

Microbiome analysis can be used to provide lifestyle information and behavioral patterns (including diet, cohabitants, pets, and romantic partners), determine the body site where a sample came from and answer contextual questions of sources of commingled body fluid samples, estimate postmortem intervals (PMI), and determine the environmental locations a body or object interacted with (Shrivastava et al. 2015, Schmedes et al. 2016, Clarke et al. 2017, Cho et al. 2019). Human microbiome analysis can enable monozygotic twin differentiation (Aly et al. 2015). It can also be used in occupational medicine investigations (Giampaoli et al. 2018) and to determine food authenticity (Arenas et al. 2017, Haynes et al. 2019).

NGS has been used to perform molecular analysis of post-mortem and autopsy samples. Analysis of post-mortem samples NGS of the 16S rRNA gene using the MiSeq could lead to the determination that a bacterial infection resulted in sepsis that caused the death (Cho et al. 2019). In a test of sixty-five post-mortem specimens in which the Illumina MiSeq and Applied Biosystems MicroSEQ 500 16S rDNA Bacteria Identification system were used for sequencing, the MiSeq was found to be more time- and cost-efficient when more than thirty samples are analyzed and was easier to use for bacterial identification with a larger library for more accurate determination (Cho et al. 2019). NGS using the Illumina platform was employed to characterize microbial species in animals that drowned as compared to those submerged postmortem (Wang et al. 2020). Unweighted UniFrac-based PCA differentiated the microbial constituency of the skin, lung, blood, and liver samples of the two groups (Wang et al. 2020). NGS was applied to determine cause of death in two recent autopsies of severe acute respiratory syndrome coronavirus 2 (SARS-CoV-2) positive individuals (Sekulic et al. 2020). Quantitative reverse transcriptase PCR was used to synthesize the cDNA for SARS-CoV-2 RNA isolated from lung tissue which was sequenced by NGS (Sekulic et al.

2020). One of the two cases "revealed mutations most consistent with Western European Clade A2a with ORF1a L3606F mutation" (Sekulic et al. 2020). This study demonstrates the power of NGS in molecular genetic analysis in cause of death analyses.

8.9 Bioinformatic Approaches and Tools

As with all NGS runs, the read quality and other quality metrics are assessed. However, unlike the human identification applications, there is no special-ized, commercial software for analyzing microbial sequence data for forensic investigations. Technical and biological validation of the applications will be required before NGS can be adopted as a standard tool for use in case-work and acceptance into courts (Kuiper 2016). The NGS data is uploaded to the cloud and analyzed using a series of open source and low cost software which uses bioinformatics to generate a taxonomic profile of the microbiota, reconstruct the microbial genomes, and perform a functional analyses of the genes. A metagenomic analysis is conducted to analyze the WGS and 16S rRNA data, determine the functional groups of the genes, and construct a taxonomic classification. Operational taxonomic units (OTUs) are generated by comparisons to a database and classified using a taxonomic classifica-tion for taxonomic profiling and used to generate a phylogenetic tree. For example, the Qiime software can be used to detect chimera, cluster OTUs, pick representative sequences, assign taxonomy, and generate a taxonomic table. MG-RAST uses the uploaded sequences and metadata to assess the quality, perform RNA identification and clustering, assign taxonomy, and prepare a taxonomy table. The mothur program can perform quality control, align sequences, clean alignments, pre-cluster sequences and perform chi-mera detection, classify sequences, remove non-bacterial sequences, generate a distance matrix, cluster and classify OTUs, and create a taxonomy table. Bar graphs are generated to demonstrate the proportion of each family of microbe present in a sample. The bar graphs look like a bar code, as shown in Figure 8.1 and differences are readily apparent.

Culture-independent analysis such as NGS involves PCR amplification followed by sequencing. Analysis tools and pipelines perform a variety of functions including data cleaning, sequence alignment, gene classification and annotation, and grouping sequences into OTUs. OTU groupings are used to infer phylogenetic and taxonomic relationships. Many analysis tools to manage and analyze NGS data were developed during the HMP time-frame. In a study, three analysis methods (QIIME, mothur, and MG-RAST) were compared for their performance in evaluating 16 S rRNA sequence data from preterm gut microbiota (Plummer et al. 2015). These tools perform tasks including quality control, sequence alignment, adding metadata, classifying sequences, nonbacterial sequence purging, RNA identification and clustering, chimera detection, OTU clustering, sequencing picking, and taxonomy and have a similar workflow (Plummer et al. 2015). Mothur was able to annotate a "slightly higher number of reads" (Plummer et al. 2015). The workflow times varied from an hour for QIIME, ten hours for mothur, and two days with manual cleaning for MG-RAST (Plummer et al. 2015). The three programs identified the sample phyla as most abundant although MG-RAST left the most phyla unclassified and failed to identify some low abundance phyla (Plummer et al. 2015). The HmmUFOtu tool processes microbiome amplicon sequencing data, clusters the data into OTUs, and assigns taxonomy (Zheng et al. 2018). In the authors' comparison to standard pipelines, HmmUFOtu was more accurate in determining microbial community diversity and composition faster with a very high accuracy (Zheng et al. 2018). For comparison, the tool had the same high accuracy of the NCBI Blastn pairwise alignment algorithm but was over 300 times faster due to its efficiency with multiple sequence alignment (Zheng et al. 2018). Other tools have also been developed to handle and clean the sequence data. As NGS produces short sequences that require mapping to a reference genome, the DecontaMiner tool was developed to detect the presence of contaminating unmapped sequences, either from the lab or the biological source, in human RNA-Seq data (Sangiovanni et al. 2019). The Kraken 2 program is a tool that offers "ultrafast and accurate 16S rRNA microbial community analysis" (Lu and Salzberg et al. 2020). The CLUSTOM tool was developed for clustering 16S rRNA NGS data by overlap minimization (Hwang et al. 2013). Identifying microbes in a majority human sample is a difficult signal-to-noise problem (Minogue et al. 2019). The PathSeq tool was shown to discover viral sequences present in human tissue sequenced by deep sequencing (Kostic et al. 2011). A full review of bioinformatics tools is beyond the scope of this chapter. However, there are many bioinformatics tools available; the one chosen for use by forensic labs will need to undergo testing validation in the lab environment prior to use on casework.

8.10 Bioforensics and Biosurveillance

NGS capabilities have developed from applications in bioforensics, biosurveillance, and infectious disease diagnosis (Budowle et al. 2014, Yang et al. 2014, Schmedes et al. 2016, Arenas et al. 2017, Minogue et al. 2019). NGS biothreat surveillance and diagnostics grew out of research funding that followed the 2001 Amerithrax case investigation (Minogue et al. 2019). The nascent tool saw little use in the Amerithrax case (Minogue et al. 2019); WGS and Sanger sequencing was used to identify the source of the specimen (Minogue et al. 2019). Broomall et al. (2016) report that sequences of gamma-irradiated mail rendering non-viable substances could be determined using the Illumina and 454 NGS platforms. NGS is being used in West Africa for Ebola virus biosurveillance (Minogue et al. 2019).

NGS can be used for source agent identification (Minogue et al. 2019). Determination of the molecular signature of a pathogen is possible with NGS (Gilchrist et al. 2015, Minogue et al. 2019). WGS coupled with NGS can be used to identify pathogenic organisms including food-borne pathogen bacteria that may be used as bioweapons and indicate bioterrorism activity in a disease outbreak (Sjödin et al. 2012, Gilchrist et al. 2015, Elkins 2019, Elkins and Bender 2020).

NGS can also be used to detect the introduction of plant and animal diseases including bacteria, virus, and fungal microorganisms for forensic applications (Sjödin et al. 2012). As an example, the presence of fungi such as *Aspergillus* can indicate the location of an indoor wall, while *Penicillium*, *Debaryomices*, and *Wickerhamomyces* indicate food storage (Giampaoli et al. 2020). The MiSeq NGS platform was used to analyze food spoilage organisms and detect that the bacteria in cooked ham was mostly from four phyla, while vacuum-packed ham samples contained three families of cold resistant flora including a *Clostridium* spp. (Piotrowska-Cyplik et al. 2016). While microorganisms may not always inform the specific environment, NGS can be used to produce sequence data that can cluster samples with similar provenance (Giampaoli et al. 2020). A metabarcoding approach was applied to forensically related environmental soil samples resulting in an accurate and sensitive analysis of organisms including microflora, plants, metazoa, and protozoa (Giampaoli et al. 2014).

8.11 Infectious Disease Diagnostics

MPS and bioinformatics characterization of microorganisms can be used to track infectious diseases (Schmedes et al. 2016). In a recent study, NGS performed using an Illumina MiSeq was used to detect emerging infectious disease from viruses and bacteria in bats using DNA extracted from blood samples (Hadi et al. 2020). The data was analyzed using Bowtie and searches using Blastn in the NCBI databank (Hadi et al. 2020). The genetic data showed that bat immunity evolved with flight ability (Hadi et al. 2020). Human genetic analysis has also shown that immunity develops with bacterial exposure (Sharma and Gilbert 2018). NGS was used instead of Sanger sequencing to investigate Hepatitis B virus (HBV) infection in ninety-four patients and forty-five chronic HBV-infected individuals; word pattern frequencies of HBV sequences differentiated the HBV genotypes and infection status (Bai et al. 2018). In another study, NGS was used to detect encephalitis virus in donated transplant organs and transmission from the organs (Lipowski et al. 2017). WGS and NGS were used to analyze *Burkholderia mallei* and *Burkholderia pseudomallei* that cause the human diseases melioidosis and glanders using SNPs and determine the direction of mutation using passaged samples (Jakupciak et al. 2013).

8.12 NGS Applications in Archeology

Metagenomics deep sequencing studies can also be applicable to investigations of ancient host microbiomes. In a limited genetic investigation, *Helicobacter pylori* was detected in ancient DNA in archeological samples from 17th-century Korean mummy stomachs (Shin et al. 2018). *H. pylori* can cause gastric disease (Shin et al. 2018). The authors identified *H. pylori* DNA vacA (s- and m-region) alleles from stomach isolates of two samples including vacA s1/m2 in a Cheongdo mummy and s1 in a Dangjin mummy and suggest that NGS is needed for full characterization of the samples (Shin et al. 2018). In another study, sequencing data was analyzed using bioinformatics methods to screen for Mycobacterium tuberculosis complex fingerprints in twenty-eight individuals (dated 4400–4000 and 3100–2900 BC) from central Poland and showed that NGS led to the identification of probable ancient disease cases supported by statistics (Borówka et al. 2019)

8.13 Summary of NGS Microbial Sequencing Applications in Forensic Investigation

Microbiome sequences contain all of the information needed to solve a variety of forensic cases. NGS using both targeted and metagenomics approaches have been applied to a wide variety of forensic applications (Figure 8.3). NGS has differentiated sources of body fluids and sites on the human body and fingerprinted their microbial composition. It has been used to determine the cause of death in an autopsy and signs of disease in oral and blood samples. It has been used to differentiate hair sources and predict lifestyle patterns. NGS has greatly enhanced the capabilities in disease diagnostics, bioforensics, and biosurveillance as well as soil profiling and geolocation. It has been used in archeological studies to identify gut bacteria and infection. Microbiome analysis will be incorporated more widely in forensic investigations in the future.

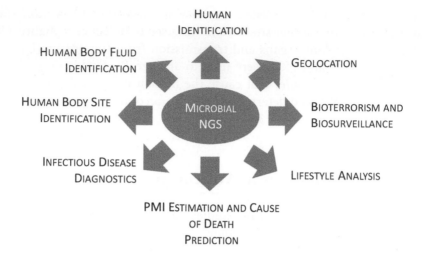

Figure 8.3 Summary of forensic applications of microbial NGS.

Questions

1. What loci are targeted in microbes for sequencing?
2. Explain how NGS is used for microbial analysis.
3. Describe data analysis approaches for cleaning and interpreting microbial DNA sequence data.
4. Explain how NGS can be used to detect and differentiate body regions a sample is derived from or was in contact with.
5. Explain how NGS of microbes can be used for geolocation.
6. List NGS issues that must be overcome with microbial DNA data analysis before the tool is widely adopted in forensic science.

References

Alessandrini, F., Brenciani, A., Fioriti, S., Melchionda, F., Mingoia, M., Morroni, G., and A. Tagliabracci. "Validation of a universal DNA extraction method for human and microbiAL DNA analysis." *Forensic Science International Genetics Supplement Series* 7, no. 1 (December 2019): 256–258. doi:10.1016/j.fsigss.2019.09.098.

Aly, S.M., and D.M. Sabri. "Next generation sequencing (NGS): a golden tool in forensic toolkit." *Archiwum medycyny sadowej i kryminologii* 65, no. 4 (2015): 260–271. doi:10.5114/amsik.2015.61029.

Arenas, M., Pereira, F., Oliveira, M., Pinto, N., Lopes, A.M., Gomes, A., Carracedo, A., and A. Amorim. "Forensic genetics and genomics: Much more than just a human affair." *PLoS Genetics* 13 (September 21, 2017): e1006960. doi:10.1371/journal.pgen.1006960.

Bai, X., Jia, J.-A., Fang, M., Chen, S., Liang, X., Zhu, S., Zhang, S., Feng, J., Sun, F., and C. Gao. "Deep sequencing of HBV pre-S region reveals high heterogeneity of HBV genotypes and associations of word pattern frequencies with HCC." *PLoS Genetics* 14, no. 2 (February 23, 2018): 1007206. doi:10.1371/journal.pgen.1007206.

Belizário, J.E., and M. Napolitano. "Human microbiomes and their roles in dysbiosis, common diseases, and novel therapeutic approaches." *Frontiers in Microbiology* 6 (October 6, 2015): 1050. doi:10.3389/fmicb.2015.01050.

Bender, A.C., Faulkner, J.A., Tulimieri, K., Boise, T.H., and K.M. Elkins. "High resolution melt assays to detect and identify *Vibrio parahaemolyticus, Bacillus cereus, Escherichia coli*, and *Clostridioides difficile* bacteria." *Microorganisms* 8, no. 4 (April 14, 2020): 561. doi:10.3390/microorganisms8040561.

Borówka, P., Puławski, L., Marciniak, B., Borowska-Strugińska, B., Dziadek, J., Żądzińska, E., Wiesław Lorkiewicz, W., and D. Strapagiel. "Screening methods for detection of ancient *Mycobacterium tuberculosis* complex fingerprints in next-generation sequencing data derived from skeletal samples." *GigaScience* 8, no. 6 (June 1, 2019): giz065. doi:10.1093/gigascience/giz065.

Broomall, S.M., Ichou, M.A., Krepps, M.D., Johnsky, L.A., Karavis, M.A., Hubbard, K.S., Insalaco, J.M., Betters, J.L., Redmond, B.W., Rivers, B.A., Liem, A.T., Hill, J.M., Fochler, E.T., Roth, P.A., Rosenzweig, C.N., Skowronski, E.W., and H.S. Gibbon. "Whole-genome sequencing in microbial forensic analysis of gamma-irradiated microbial materials." *Applied Environmental Microbiology* 82, no. 2 (January 2016): 596–607. doi:10.1128/AEM.02231-15.

Budowle, B., Connell, N.D., Bielecka-Oder, A., Colwell, R.R., Corbett, C.R., Fletcher, J., Forsman, M., Kadavy, D.R., Markotic, A., Morse, S.A., Murch, R.S., Sajantila, A., Schmedes, S.E., Ternus, K.L., Turner, S.D., and S. Minot. "Validation of high throughput sequencing and microbial forensics applications." *Investigative Genetics* 5 (June 30, 2014): 9. doi:10.1186/2041-2223-5-9.

Chen, W., Cheng, Y.M., Zhang, S.W., and Q. Pan. "Supervised method for periodontitis phenotypes prediction based on microbial composition using 16S rRNA sequences." *International Journal of Computational Biology and Drug Design* 7, no. 2 (May 28, 2014): 214–224. doi:10.1504/IJCBDD.2014.061647.

Cho, Y., Lee, M.H., Kim, H.S., Park, M., Kim, M.-H., Kwon, H., Kim, J., Lee, Y.H., and D.S. Lee. "Comparative analysis of Sanger and next generation sequencing methods for 16S rDNA analysis of post-mortem specimens." *Australian Journal of Forensic Sciences* 51, no. 5 (2019): 426–455. doi:10.1080/00450618.2017.1402957.

Chowdhury, S., and S.S. Fong. "Computational modeling of the human microbiome." *Microorganisms* 8, no. 2 (January 21, 2020): 197. doi:10.3390/microorganisms 8020197.

Clarke, T.H., Gomez, A., Singh, H., Nelson, K.E., and L.M. Brinkac. "Integrating the microbiome as a resource in the forensics toolkit." *Forensic Science International: Genetics* 30 (September 2017): 141–147. doi:10.1016/j.fsigen.2017.06.008.

Elkins, K.M. *Introduction to Forensic Chemistry*. Boca Raton, FL: CRC Press/ Taylor & Francis, 2019.

Elkins, K., and A. Bender. "Detection and identification of foodborne pathogens." *Encyclopedia* 1 (2020). https://encyclopedia.pub/512.

Franco-Duarte, R., Černáková, L., Snehal, K., Kaushik, K.S., Salehi, B., Bevilacqua, A., Corbo, M.R., Antolak, H., Dybka-Stępień, K., Leszczewicz, M., and S.

Relison Tintino. "Advances in chemical and biological methods to identify microorganisms—From past to present." *Microorganisms* 7, no. 5 (May 13, 2019): 130. doi:10.3390/microorganisms7050130.

Giampaoli, S., Berti, A., Di Maggio, R.M., Pilli E., Valentini, A., Valeriani, F., Gianfranceschi, G., Barni F., Ripani, L., and V.R. Spica. "The environmental biological signature: NGS profiling for forensic comparison of soils." *Forensic Science International* 240 (July 2014): 41–47. doi:10.1016/j.forsciint.2014.02.028.

Giampaoli, S., DeVittori, E., Valeriani, F., Berti, A., and V. Romano Spica. "Informativeness of NGS analysis for vaginal fluid identification." *Journal of Forensic Sciences* 62, no. 1 (January 2017): 192–196. doi:10.1111/1556-4029.13222.

Giampaoli, S., Alessandrini, F., Frajese, G.V., Guglielmi, G., Tagliabracci, A., and A. Berti. "Environmental microbiology: Perspectives for legal and occupational medicine." *Legal Medicine (Tokyo)* 35 (November 2018): 34–43. doi:10.1016/j.legalmed.2018.09.014.

Giampaoli, S., De Vittori E., Frajese, G.V., Paytuví, A., Sanseverino, W, Anselmo, A., Barni, F., and A. Berti. "A semi-automated protocol for NGS metabarcoding and fungal analysis in forensic." *Forensic Science International* 306 (January 2020): 110052. doi:10.1016/j.forsciint.2019.110052.

Gilchrist, C.A., Turner, S.D., Riley, M.F., Petri, W.A., and E.L. Hewlett. "Whole-genome sequencing in outbreak analysis." *Clinical Microbiology Reviews* 28, no. 3 (July 2015): 541–563. doi:10.1128/CMR.00075-13.

Gurenlian, J.R. "The role of dental plaque biofilm in oral health." *Journal of Dental Hygiene* 81, no. 5 (October 2007): 116. https://jdh.adha.org/content/jdenthyg/81/suppl_1/116.full.pdf.

Hadi, M.I., Alamudi, M.Y., Suprayogi, D., and M. Widiyanti. "Detection of emerging infectious disease in *Chiroptera brachjatis* and *Rhinolopus boorneensis* as reservoirs of zoonotic diseases in Indonesia." *Indian Journal of Forensic Medicine and Toxicology* 14(2020): 2027–2032.

Hasan, N.A., Young, B.A., Minard-Smith, A.T., Saeed, K., Li, H., Heizer, E.M., McMillan, N.J., Isom, R., Abdullah, A.S., and D.M. Bornman. "Microbial community profiling of human saliva using shotgun metagenomic sequencing." *PLoS One* 9 (May 20, 2014): e97699. doi:10.1371/journal.pone.0097699.

Haynes, E., Jimenez, E., Pardo, M.A., and S.J. Helyar. "The future of NGS (Next Generation Sequencing) analysis in testing food authenticity." *Food Control* 101 (July 2019): 134–143. doi:10.1016/j.foodcont.2019.02.010.

Hwang, K., Oh, J., Kim, T.K., Kim, B.K., Yu, D.S., Hou, B.K., Caetano-Anollés, G., Hong, S.G., and K.M. Kim. "CLUSTOM: a novel method for clustering 16S rRNA next generation sequences by overlap minimization." *PLoS One* 8, no. 5 (May 1, 2013): e62623. doi:10.1371/journal.pone.0062623.

Jakupciak, J.P., Wells, J.M., Karalus, R.J., Pawlowski, D.R., Lin, J.S., and A.B. Feldman. "Population-sequencing as a biomarker of *Burkholderia mallei* and *Burkholderia pseudomallei* evolution through microbial forensic analysis." *Journal of Nucleic Acids* 2013 (December 17, 2013): 801505. doi:10.1155/2013/801505.

Kistler, J.O., Pesaro, M., and W.G. Wade. "Development and pyrosequencing analysis of an in-vitro oral biofilm model." *BMC Microbiology* 15 (February 10, 2015): 24. doi:10.1186/s12866-015-0364-1.

Kostic, A.D., Ojesina, A., Pedamallu, C.S., Jung, J., Verhaak, R.G.W., Getz, G., and M. Meyerson. "PathSeq: Software to identify or discover microbes by deep sequencing of human tissue." *Nature Biotechnology* 29, no. 5 (May 2011): 393–396. doi:10.1038/nbt.1868.

Kuiper, I. "Microbial forensics: next-generation sequencing as catalyst: The use of new sequencing technologies to analyze whole microbial communities could become a powerful tool for forensic and criminal investigations." *EMBO Reports* 17 (2016): 1085–1087. doi:10.15252/embr.201642794..

Leong, L.E.X., Taylor, S.L, Shivasami, A., Goldwater, P.N., and G.B. Rogers. "Intestinal microbiota composition in sudden infant death syndrome and age-matched controls." *Journal of Pediatrics* 191 (December 2017): 63–68. doi:10.1016/j.jpeds.2017.08.070.

Lilje, L., Lillsaar, T., Rätsep, R., Simm, J., and A. Aaspõllu. "Soil sample metagenome NGS data management for forensic investigation." *Forensic Science International: Genetics Supplement Series* 4, no. 1 (2013): e35–e36. doi:10.1016/j.fsigss.2013.10.017.

Lipowski, D., Popiel, M., Perlejewski, K., Nakamura, S., Bukowska-Osko, I., Rzadkiewicz, E., Dzieciatkowski, T., Milecka, A., Wenski, W., Ciszek, M., and A. Dębska-Ślizień. "A cluster of fatal tick-borne encephalitis virus infection in organ transplant setting." *The Journal of Infectious Diseases* 215, no. 6 (March 15, 2017): 896–901. doi:10.1093/infdis/jix040.

Lu, J., and S.L. Salzberg. "Ultrafast and accurate 16S rRNA microbial community analysis using Kraken 2." *Microbiome* 8 (2020): 124. doi:10.1186/s40168-020-00900-2.

Milani, C., Duranti, S., Bottacini, F., Casey, E., Turroni, F., Mahony, J., Belzer, C., Palacio, S.D., Montes, S.A., Mancabelli, L., and G.A. Lugli. "The first microbial colonizers of the human gut: Composition, activities, and health implications of the infant gut microbiota." *Microbiology and Molecular Biology Reviews* 81, no. 4 (November 8, 2017): e00036-17. doi:10.1128/MMBR.00036-17.

Minogue, T.D., Koehler, J.W., Stefan, C.P., and T.A. Conrad. "Next-generation sequencing for biodefense: Biothreat detection, forensics, and the clinic." *Clinical Chemistry* 65, no. 3 (March 1, 2019): 383–392. doi:10.1373/clinchem.2016.266536.

Mitra, S., Förster-Fromme, K., Damms-Machado, A., Scheurenbrand, T., Biskup, S., Huson, D. H., and S.C. Bischoff. "Analysis of the intestinal microbiota using SOLiD 16S rRNA gene sequencing and SOLiD shotgun sequencing." *BMC Genomics* 14, Suppl no. 5 (October 16, 2013): S16. doi:10.1186/1471-2164-14-S5-S16.

Pasolli, E., Asnicar, F., Manara, S., Zolfo, M., Karcher, N., Armanini, F., Begini, F., Manghi, P., Tett, A., Ghensi, P., Collado, M.C., Rice, B.L., DuLong, C., Morgan, X.C., Golden, C.D., Quince, C., Huttenhower, C., and N. Segata. "Extensive unexplored human microbiome diversity revealed by over 150,000 genomes from metagenomes spanning age, geography, and lifestyle." *Cell* 176, no. 3 (January 24, 2019): 649–662. doi:10.1016/j.cell.2019.01.001.

Phadke, S., Salvador, A.F., Alves, J.I., Bretschger, O., Alves, M.M., and M.A. Pereira. "Harnessing the power of PCR molecular fingerprinting methods and next generation sequencing for understanding structure and function in microbial communities." In *PCR*, 1st ed., edited by L. Domingues, 225–248. New York: Springer, 2017. doi:10.1007/978-1-4939-7060-5_16.

Piotrowska-Cyplik, A., Myszka, K., Czarny, J., Ratajczak, K., Kowalski, R., and R. Biegańska-Marecik. "Characterization of specific spoilage organisms (SSOs) in vacuum-packed ham by culture-plating techniques and MiSeq next-generation sequencing technologies." *Journal of the Science of Food and Agriculture* 97 (January 2016): 659–668. doi:10.1002/jsfa.7785.

Plummer, E., Twin, J., Bulach, D.M., Garland, S.M., and S.N. Tabrizi. "A comparison of three bioinformatics pipelines for the analysis of preterm gut microbiota using 16S rRNA gene sequencing data." *Journal of Proteomics and Bioinformatics* 8, no. 12 (December 2015). doi:12. 10.4172/jpb.1000381.

Quaak, F.C.A., van Duijn, T., Hoogenboom, J., Kloosterman, A.D., and I. Kuiper. "Human-associated microbial populations as evidence in forensic casework." *Forensic Science International: Genetics* 36 (September 2018): 176–185. doi:10.1016/j.fsigen.2018.06.020.

Rajan, S.K., Lindqvist, M., Brummer, R.J., Schoultz, I., and D. Repsilber. "Phylogenetic microbiota profiling in fecal samples depends on combination of sequencing depth and choice of NGS analysis method." *PLoS One* 14, no. 9 (September 2019): e0222171. doi:10.1371/journal.pone.0222171.

Sangiovanni, M., Granata, I., Thind, A.S., and M.R. Guarracino. "From trash to treasure: Detecting unexpected contamination in unmapped NGS data." *BMC Bioinformatics* 20, Suppl 4 (April 18, 2019): 168. doi:10.1186/s12859-019-2684-x.

Schmedes, S.E., Sajantila, A., and B. Budowle. "Expansion of microbial forensics." *Journal of Clinical Microbiology* 54 (2016): 1964–1974. doi:10.1128/JCM.00046-16.

Schmedes, S.E., Woerner, A.E., Novroski, N.M.M., Wendt, F.R., King, J.L., Stephens, K.M., and B. Budowle. "Targeted sequencing of clade-specific markers from skin microbiomes for forensic human identification." *Forensic Science International: Genetics* 32 (January 2018): 50–61. doi:10.1016/j.fsigen.2017.10.004.

Sekulic, M., Harper, H., Nezami, B.G., Shen, D.L, Sekulic, S.P., Koeth, A.T., Harding, C.V., Gilmore, H., and N. Sadri. "Molecular detection of SARS-CoV-2 infection in FFPE samples and histopathologic findings in fatal SARS-CoV-2 cases." *American Journal of Clinical Pathology* 154, no. 2 (July 7, 2020): 190–200. doi:10.1093/ajcp/aqaa091.

Sharma, A., and J.A. Gilbert. "Microbial exposure and human health." *Current Opinion in Microbiology* 44 (August 2018): 79–87. doi:10.1016/j.mib.2018.08.003.

Shin, D.H,. Oh, C.S., Hong, J.H., Lee, H., Lee, S.D., and E. Lee. "*Helicobacter pylori* DNA obtained from the stomach specimens of two 17(th) century Korean mummies." *Anthropologischer Anzeiger* 75, no. 1 (2018): 75–87. doi:10.1127/anthranz/2018/0780.

Shrivastava, P., Jain, T., and M.K. Gupta. "Microbial forensics in legal medicine." *SAS Journal of Medicine* 1 (2015): 33–40.

Siqueira, J.F., Fouad, A.F., and I.N. Rôças. "Pyrosequencing as a tool for better understanding of human microbiomes." *Journal of Oral Microbiology* 4, no. 1 (2012): 10743. doi:10.3402/jom.v4i0.10743.

Sjödin, A., Broman, T., Melefors, O., Andersson, G., Rasmusson, B., Knutsson, R., and M. Forsman. "The need for high-quality whole-genome sequence databases in microbial forensics." *Biosecurity and Bioterrorism* 11 (2012): S78–S86. doi:10.1089/bsp.2013.0007.

Tozzo, P., D'Angiolella, G., Brun, P., Castagliuolo, I., Gino, S., and L. Caenazzo. "Skin microbiome analysis for forensic human identification: What do we know so far?" *Microorganisms* 8 (June 9, 2020): 873. doi:10.3390/microorganisms8060873.

Tridico, S.R., Murray, D.C., Addison, J., Kirkbride, K.P., and M. Bunce. "Metagenomic analyses of bacteria on human hairs: A qualitative assessment for applications in forensic science." *Investigative Genetetics* 5 (December 16, 2014): 16. doi:10.1186/s13323-014-0016-5.

Wang, L.-L. Zhang, F.-Y., Dong, W.-W., Wang, C.-L., Liang, X.-Y., Suo, L.-L., Cheng, J., Zhang, M., Guo, X.-S., Jiang, P.-H., Guan, D.-W., and R. Zhao. "A novel approach for the forensic diagnosis of drowning by microbiological analysis with next-generation sequencing and unweighted UniFrac-based PCoA." *International Journal of Legal Medicine* 134, no. 6 (2020): 2149–2159. doi:10.1007/s00414-020-02358-1.

Welinder-Olsson, C., Dotevall, L., Hogevik, H., Jungnelius, R., Trollfors, B., Wahl, M., and P. Larsson. "Comparison of broad-range bacterial PCR and culture of cerebrospinal fluid for diagnosis of community-acquired bacterial meningitis." *Clinical Microbiology and Infection* 13, no. 9 (September 2007): 879–886. doi:10.1111/j.1469-0691.2007.01756.x.

Willis, J.R., and T. Gabaldón. "The human oral microbiome in health and disease: From sequences to ecosystems." *Microorganisms* 8, no. 2 (February 23, 2020): 308. doi:10.3390/microorganisms8020308.

Woerner, A.E., Novroski, N.M.M., Wendt, F.R., Ambers, A., Wiley, R., Schmedes, S.E., and B. Budowle. "Forensic human identification with targeted microbiome markers using nearest neighbor classification." *Forensic Science International: Genetics* 38 (January 2019): 130–139. doi:10.1016/j.fsigen.2018.10.003.

Wojciuk, B., Salabura, A., Grygorcewicz, B., Kędzierska, K., Ciechanowski, K., and B. Dołęgowska. "Urobiome: In sickness and in health." *Microorganisms* 7 (November 10, 2019): 548. doi:10.3390/microorganisms7110548.

Yang, Y., Xie, B., and J. Yan. "Application of next-generation sequencing technology in forensic science." *Genomics, Proteomics & Bioinformatics* 12, no. 5 (October 2014): 190–197. doi:10.1016/j.gpb.2014.09.001.

Zheng, Q., Bartow-McKenney, C., Meisel, J.S., and E.A. Grice. "HmmUFOtu: An HMM and phylogenetic placement based ultra-fast taxonomic assignment and OTU picking tool for microbiome amplicon sequencing studies." *Genome Biology* 19 (June 27, 2018): 82. doi:10.1186/s13059-018-1450-0.

Body Fluid Analysis Using Next Generation Sequencing

9

9.1 Introduction

Body fluids analysis is performed in forensic cases to understand the circumstances of the case. Traditional methods of body fluid analysis include colorimetric, enzymatic, immunochromatographic, enzyme-linked immunosorbent assay (ELISA), and microscopy assays (Virkler and Lednev 2009). While these have many strengths, specifically low cost, ease of use, and portability, there is room for improvement. For example, with the current methods, the analyst must decide which test to perform based on their expectations of which body fluid is likely to be present and then each body fluid must be tested individually. This is a significant drawback as casework evidence samples may contain more than one body fluid from one or more donors. In addition, many of the traditional tests require intact enzymes which depend on the stability of these molecules. Conditions that lead to damaged and degraded samples often degrade or render enzyme molecules inactive. Additionally, the sensitivity of the tests varies, and some are not very sensitive. Similarly, many of the tests are not specific and have known false positives. Furthermore, trace evidence is submitted in many cases and is simply not sufficient to permit both body fluid analysis and identity testing. Finally, traditional body fluid tests are destructive, and the quantity of the evidence sample may preclude body fluid testing if identity testing is to be performed. Next generation sequencing (NGS) offers a solution to several of the above problems. NGS methods including pyrosequencing and massively parallel sequencing have been developed for multiplex body fluid testing. Samples include mRNA and methylated DNA that can be co-extracted during sample preparation steps for human identification testing. NGS is not without its own drawbacks. It requires specialized instrumentation and analyst training. Nevertheless, it offers high specificity and sensitivity, simultaneous identification of body fluids, and reproducibility.

9.2 Epigenetic-Based Tissue Source Attribution

Epigenetics is the study of gene expression. Gene expression varies by cell type but can also change with environmental factors, aging, and disease such

DOI: 10.4324/9781003196464-9   137

as carcinogenesis (Lilischkis et al. 2001, Bernstein et al. 2007). Gene expression can be modulated by chemical modification to the cytosine nitrogenous base in DNA. Select cytosine residues are methylated at C-5 of the pyrimidine base and are termed 5′-methylcytosine (Lilischkis et al. 2001, Bernstein et al. 2007). The methylation of cytosine creates an additional layer for individualization of DNA valuable for identifying identical twins. Cytosine methylation in DNA was found to be a heritable trait like the sequence of the nitrogenous bases in the human genome, and methylation sites have been annotated (Meissner et al. 2005, Bernstein et al. 2007, Harrow et al. 2012). Methylated cytosines generally precede guanine bases forming so-called CpG islands in the 5′ to 3′ direction on the DNA strand (Lilischkis et al. 2001, Bernstein et al. 2007). CpG islands represent approximately 2% of the cytosines in the human genome and are most commonly found in the promoter region of genes (Lilischkis et al. 2001). While they usually repress transcription, in some cases they promote transcription (Lilischkis et al. 2001, Weber et al. 2005). Methylated cytosines cluster in CpG islands, but the pattern of CpG methylation varies among body fluids and tissues (Lilischkis et al. 2001). Paliwal et al. (2010) reviewed quantitative detection of DNA methylation at CpG sites and developed a quantitation method for minute DNA.

Pyrosequencing is an NGS tool that has been used to identify and quantify differentially methylated loci (Dejeux et al. 2009, Powell et al. 2018). Pyrosequencing theory is covered in Chapter 2. Prior to sequencing, the extracted DNA can be treated with bisulfite which, at low pH, converts the unmethylated cytosines in the sequence to uracil (Grunau et al. 2001, Lilischkis et al. 2001). Using PCR, the target is amplified, and the uracil bases are replaced with thymine in the amplified DNA. Methylated cytosines are copied as cytosines in PCR. Tissue-specific differentially methylated regions (tDMRs) have been identified and serve as the targets in forensic body tissue and fluid analysis (Rakyan et al. 2008, Frumkin et al. 2011, Lee et al. 2012, Slieker et al. 2013, Balamuragan et al. 2014). Genomic DNA that was extracted from 3-mm dried blood spots, bisulfite treated, and PCR-amplified was sequenced to analyze the imprinted gene SNRPN (Xu et al. 2012). Slieker et al. (2013) used an Illumina 450k chip to identify and annotate tDMRs for blood, saliva, buccal swabs, and hair follicle tissue and found that most tDMRs were in CpG-poor regions. DNA extracts from evidence can be fully, partially, or unmethylated at the sites of interest. The output pyrogram produced in pyrosequencing can be used to quantify the percent methylation (Dejeux et al. 2009, Qiagen PyroMark® Q48 Autoprep User Manual, 2020).

Park et al. (2014) evaluated over 450,000 CpG sites using an Illumina 450k chip. They identified eight markers that demonstrated high sensitivity and specificity for body fluid identification (Park et al. 2014). Bruce McCord and his group has developed several pyrosequencing assays for body fluids

Table 9.1   Genetic Markers Identified Body Fluid Identification Using Pyrosequencing

Marker	Body Fluid
AHRR, cg06379435, C20orf117, cg06379435, cg08792630	Blood
ZC3H12D, FGF7, cg23521140, cg17610929	Semen
NMUR2, UBE2U, B_SPTB_03	Sperm
BCAS4, SA-6, cg26107890, cg20691722	Saliva/oral epithelia
PFN3A, VE_8	Vaginal epithelia
cg01774894, cg14991487	Vaginal secretions

including blood, saliva, semen, and vaginal fluid using the Qiagen PyroMark Assay Design software using various marker sites for the body fluids (Table 9.1). Elkins has described details of pyrosequencing primer design for CpG targets (Elkins 2021), and McCord's group has published several reports over the past half dozen years detailing the sequences and applications of pyrosequencing primers they have designed and tested for body fluid analysis (Madi et al. 2012, Balamurugan et al. 2014, Antunes et al. 2016, Silva et al. 2016, Gauthier et al. 2019, Alghanim et al. 2020). His team has reported a pyrosequencing multiplex assay that detects and identifies blood, saliva, semen, and vaginal cells simultaneously that has been licensed and distributed by Qiagen (Powell et al. 2018). In the assay, BCAS4, ZC3H12D, cg06379435, and VE_8 loci are used to detect and identify saliva, semen, blood, and vaginal epithelial cells, respectively, after they have been bisulfite treated using the Qiagen EpiTect® Fast DNA Bisulfite Kit (Figure 9.1).

9.3 mRNA-Based Tissue Source Attribution

In addition to methylated DNA targets, RNA targets have also been examined for forensic body fluid analysis. Many RNA molecules have been characterized including messenger RNA (mRNA), microRNA (miRNA), and small nuclear RNA (snRNA). Several studies have evaluated NGS approaches for differentiating body fluids using mRNA. In 2015, Børsting and Morling reviewed forensic applications of NGS using markers such as mRNA. Zubakov et al. (2015) evaluated the Ion Torrent PGM instrument in its ability to simultaneously individualize samples by STR DNA typing, perform sex typing by amelogenin, and perform body fluid/tissue identification. Dørum et al. (2018) analyzed 183 body fluids/tissues using their NGS mRNA approach using MPS and used partial least squares (PLS) and linear discriminant analysis (LDA) to classify samples using NGS read counts into one of six body fluids. They tested the model on mixed body fluid samples and its ability to identify the

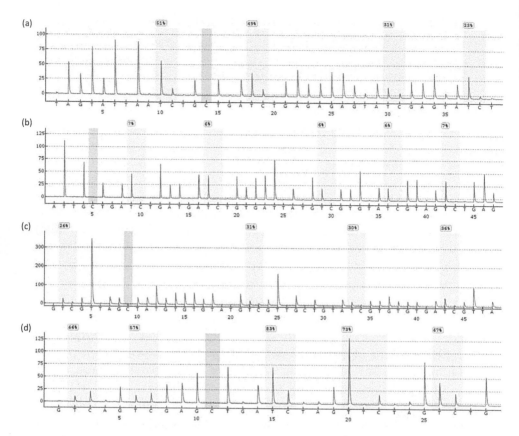

Figure 9.1 Pyrograms resulting from a vaginal epithelial sample analyzed with the Body Fluid Identification Multiplex. Vaginal epithelia is characterized by moderate methylation in the BCAS4 assay (a), hypomethylation in the cg06379435 assay (b), moderate methylation in the VE_8 assay (c), and hypermethylation in the ZC3H12D assay (d). The combination of multiple body fluid assays in a single reaction allows for higher accuracy in body fluid identification while reducing sample consumption and costs. (Courtesy of Quentin Gauthier.)

individual body fluid components in a mixture (Dørum et al. 2018). In 2018, Hanson et al. published a study of their work using targeted mRNA sequencing to identify blood, saliva, semen, vaginal secretions, menstrual blood, and skin using a thirty-three-biomarker assay. Their assay correctly identified the body fluids in a study in their lab and in a blinded study (Hanson et al. 2018).

9.4 MicroRNA Analysis

MicroRNA (miRNA) was proposed for forensic applications by Courts and Madea in 2010. At 18–24 nucleotides in length (Yang et al. 2014), miRNA molecules are shorter in length than mRNA and small nuclear RNA (snRNA) molecules. Like mRNA and snRNA, they are endogenous to the organism (Hanson et al. 2009, Yang et al. 2014). Their small size resists degradation (Yang et al. 2014). Thus, miRNA profiling is an attractive tool for typing damaged or compromised samples. Hanson et al. (2009) analyzed 452 miRNAs from different body fluids including blood, menstrual blood, semen, saliva, and vaginal secretions and observed differential expression in nine miRNAs (miR451, miR16, miR135b, miR10b, miR658, miR205, miR124a, miR372, and miR412) using real-time PCR and demonstrated the use of miRNA for forensic applications using as little as 50 pg of sample. In 2010, Zubakov et al. (2010) analyzed the expression of 718 miRNAs from saliva, semen, venous blood, menstrual blood, and vaginal secretions using a microarray and evaluated using RT-PCR Taq Man assays for distinguishing the body fluids and reported miRNAs specific for venous blood and semen. NGS technology was proposed for miRNA forensic analysis in 2014 (Yang et al. 2014); however, at the time of this writing, no researchers have published initial proof-of-concept studies.

9.5 The Future of Body Fluid Assays

Massively parallel sequencing (MPS) using the MiSeq and Ion Torrent instruments also offers new opportunities for body fluid analysis. Just as PCR primers have been designed, tested, and multiplexed for human identification and phenotyping applications, they can be designed to target loci such as SNPs with demonstrated variations in body fluids. Alternatively, extracted snRNA or miRNA could be reverse transcribed to cDNA and the targets could be sequenced using MPS using sequencing as is being performed with mRNA. However, as with the introduction of NGS applications for human identity testing, labs need to allocate time and resources to acquire the instrumentation, consumables, and training and validate the new methods. The commercialization of the pyrosequencing assays for body fluid analysis is an indication of the importance of this capability. How widely the new tool is adopted will depend not only on funding but also institutional desire and will to make a significant change in evidence processing and standard operating procedure.

Questions

1. List a locus for the identification of each body fluid.
2. Explain how mRNA, microRNA, and methylated DNA can be used in body fluid analysis.

3. What NGS technique has been developed to analyze body fluids? Explain how it works.
4. Could the MiSeq FGx or Ion Torrent instruments be used in body fluid analysis? Explain your answer.
5. What is the future of body fluid testing? Explain.

References

Alghanim, H., Balamurugan, K., and B. McCord. "Development of DNA methylation markers for sperm, saliva and blood identification using pyrosequencing and qPCR/HRM." *Analytical Biochemistry* 611 (December 15, 2020): 113933. doi:10.1016/j.ab.2020.113933.

Antunes, J., Silva, D.S., Balamurugan, K., Duncan, G., Alho, C.S., and B. McCord. "Forensic discrimination of vaginal epithelia by DNA methylation analysis through pyrosequencing." *Electrophoresis* 37, no. 21 (October 2016): 2751–2758. doi:10.1002/elps.201600037.

Balamurugan, K., Bombardi, R., Duncan, G., and B. McCord. "Identification of spermatozoa by tissue-specific differential DNA methylation using bisulfite modification and pyrosequencing." *Electrophoresis* 35, no. 21–22 (November 2014): 3079–3086.

Bernstein, B.E., Meissner, A., and E.S. Lander. "The mammalian epigenome." *Cell* 128, no. 4, (February 23, 2007): 669–681. doi:10.1016/j.cell.2007.01.033.

Børsting, C., and N. Morling. "Next generation sequencing and its applications in forensic genetics." *Forensic Science International Genetics* 18 (September 2015): 78–89. doi:10.1016/j.fsigen.2015.02.002.

Courts, C., and B. Madea. "Micro-RNA – A potential for forensic science?" *Forensic Science International* 203, no. 1–3 (November 1, 2010): 106–111. doi:10.1016/j.forsciint.2010.07.002.

Dejeux, E., El Abdalaoui, H., Gut, I.G., and J. Tost. "Identification and quantification of differentially methylated loci by pyrosequencing™ technology." In *Methods in Molecular Biology: DNA Methylation: Methods and Protocols*, 2nd ed., vol. 507, edited by J. Tost, 189–205. New York: Humana Press, 2009. doi:10.100 7/978-1-59745-522-0_15.

Dørum, G., Ingold, S., Hanson, E., Ballantyne, J., Snipen, L., and C. Haas. "Predicting the origin of stains from next generation sequencing mRNA data." *Forensic Science International Genetics* 34 (May 2018): 37–48. doi:10.1016/j.fsigen.2018.01.001.

Elkins, K.M. "Pyrosequencing primer design for forensic biology applications." In *Methods in Molecular Biology: PCR Primer Design*, 3rd ed., edited by C. Basu. New York: Humana Press, 2021, in press.

Frumkin, D., Wasserstrom, A., Budowle, B., and A. Davidson. "DNA methylation-based forensic tissue identification." *Forensic Science International: Genetics* 5, no. 5 (November 2011): 517–524. doi:10.1016/j.fsigen.2010.12.001.

Gauthier, Q.T., Cho, S., Carmel, J.H., and B.R. McCord. "Development of a body fluid identification multiplex via DNA methylation analysis." *Electrophoresis* 40, no. 18–19 (September 2019): 2565–2574. doi:10.1002/elps.201900118.

Grunau, C., Clark, S.J., and A. Rosenthal. "Bisulfite genomic sequencing: systematic investigation of critical experimental parameters." *Nucleic Acids Research* 29, no. 13 (July 1, 2001): E65–E65. doi:10.1093/nar/29.13.e65.

Hanson, E., Ingold, S., Haas, C., and J. Ballantyne. "Messenger RNA biomarker signatures for forensic body fluid identification revealed by targeted RNA sequencing." *Forensic Science International: Genetics* 34 (May 2018): 206–221. doi:10.1016/j.fsigen.2018.02.020.

Hanson, E.K., Lubenow, H., and J. Ballantyne. "Identification of forensically relevant body fluids using a panel of differentially expressed microRNAs." *Analytical Biochemistry* 387, no. 2 (April 15, 2009): 303–314. doi:10.1016/j.ab.2009.01.037.

Harrow, J., Frankish, A., Gonzalez, J.M., Tapanari, E., Diekhans, M., Kokocinski, F., Aken, B.L., Barrell, D., Zadissa, A., Searle, S., Barnes, I., Bignell, A., Boychenko, V., Hunt, T., Kay, M., Mukherjee, G., Rajan, J., Despacio-Reyes, G., Saunders, G., Steward, C., Harte, R., Lin, M., Howald, C., Tanzer, A., Derrien, T., Chrast, J., Walters, N., Balasubramanian, S., Pei, B., Tress, M., Rodriguez, J.M., Ezkurdia, I., van Baren, J., Brent, M., Haussler, D., Kellis, M., Valencia, A., Reymond, A., Gerstein, M., Guigó, R., and T.J. Hubbard. "GENCODE: the reference human genome annotation for The ENCODE Project." *Genome Research* 22, no. 9 (September 2012): 1760–1774. doi:10.1101/gr.135350.111.

Lee, H.Y., Park, M.J., Choi, A., An, J.H., Yang, W.I., and K.-J. Shin. "Potential forensic application of DNA methylation profiling to body fluid identification." *International Journal of Legal Medicine* 126(2012): 55–62.

Lilischkis, R., Kneitz, H., and H. Kreipe. "Methylation analysis of CpG islands." In *Methods in Molecular Medicine: Metastasis Research Protocols, Volume I: Cells and Tissues*, Vol. 57, edited by S.A. Brooks and U. Schumacher, 271–283. New York: Humana Press, 2001. doi:10.1385/1-59259-136-1:271.

Madi, T., Balamurugan, K., Bombardi, R., Duncan, G., and B. McCord. "The determination of tissue-specific DNA methylation patterns in forensic biofluids using bisulfite modification and pyrosequencing." *Electrophoresis* 33 (2012): 1736–1745. doi:10.1002/elps.201100711.

Meissner, A., Gnirke, A., Bell, G.W., Ramsahoye, B., Lander, E.S., and R. Jaenisch. "Reduced representation bisulfite sequencing for comparative high-resolution DNA methylation analysis." *Nucleic Acids Research* 33, no. 18 (October 13, 2005): 5868–5877. doi:10.1093/nar/gki901.

Paliwal, A., Vaissiere, T., and Z. Herceg. "Quantitative detection of DNA methylation states in minute amounts of DNA from body fluids." *Methods* 52, no. 3 (November 2010): 242–247. doi:10.1016/j.ymeth.2010.03.008.

Park, J.-L., Kwon, O.-H., Kim, J.H., Yoo, H.-S., Lee, H.-C., Woo, K.-M., Kim, S.-Y., Lee, S.-H., and Y.S. Kim. "Identification of body fluid-specific DNA methylation markers for use in forensic science." *Forensic Science International Genetics* 13 (2014): 147–153. doi:10.1016/j.fsigen.2014.07.011.

Powell, M., Lee, A.S., St. Andre, P., and B. McCord. "Tissue source attribution using the PyroMark® Q48 Autoprep System: Sperm identification in forensic casework. Qiagen Applications Note." (2018). https://www.qiagen.com/us/resources/download.aspx?id=ddaa262e-f3ec-4ac7-9bac-aaf3ec8968cf&lang=en.

Qiagen. "PyroMark Assay Design SW 2.0 quick-start guide - (EN)." Accessed January 11, 2021. https://www.qiagen.com/us/resources/download.aspx?id=231a6894-57d4-4f0b-81f2-eaa56a2b6bd8&lang=en.

Qiagen. "PyroMark® Q48 Autoprep User Manual, June 2020." Accessed January 11, 2021. https://www.qiagen.com/us/resources/download.aspx?id=650a0c13-3b8e-4a77-b433-6b1e50b9525a&lang=en.

Rakyan, V.K., Down, T.A., Thorne, N.P., Flicek, P., Kulesha, E., Graf, S., Tomazou, E.M., Bäckdahl, L., Johnson, N., Herberth, M., Howe, K.L., Jackson, D.K., Miretti, M.M., Fiegler, H., Marioni, J.C., Birney, E., Hubbard, T.J., Carter, N.P., Tavaré, S., and S. Beck. "An integrated resource for genome-wide identification and analysis of human tissue-specific differentially methylated regions (tDMRs)." *Genome Research* 18, no. 9 (June 23, 2008): 1518–1529. doi:10.1101/gr.077479.108.

Silva, D.S.B.S., Antunes, J., Balamurugan, K., Duncan, G., Alho, C.S., and B. McCord. "Developmental validation studies of epigenetic DNA methylation markers for the detection of blood, semen and saliva samples." *Forensic Science International Genetics* 23 (July 2016): 55–63. doi:10.1016/j.fsigen.2016.01.017.

Slieker, R.C., Bos, S.D., Goeman, J.J., Bovée, J.V., Talens, R.P., van der Breggen, R., Suchiman, H.E.D., Lameijer, E.-W., Putter, H., van den Akker, E.B., Zhang, Y., Jukema, J.W., Slagboom, P.E., Meulenbelt, I., and B.T. Heijmans. "Identification and systematic annotation of tissue-specific differentially methylated regions using the Illumina 450k array." *Epigenetics Chromatin* 6, no. 1 (August 6, 2013): 26. doi:10.1186/1756–8935–6–26.

Virkler, K., and I.K. Lednev. "Analysis of body fluids for forensic purposes: From laboratory testing to non-destructive rapid confirmatory identification at a crime scene." *Forensic Science International* 188, no. 1–3 (March 2009): 1–17. doi:10.1016/j.forsciint.2009.02.013.

Weber, M., Davies, J.J., Wittig, D., Oakeley, E.J., Haase, M., and W.L. Lam. "Chromosome-wide and promoter-specific analyses identify sites of differential DNA methylation in normal and transformed human cells." *Nature Genetics* 37, no. 8 (August 2005): 853–862. doi:10.1038/ng1598.

Xu, H., Zhao, Y., Liu, Z., Zhu, W., Zhou, Y., and Z. Zhao. "Bisulfite genomic sequencing of DNA from dried blood spot microvolume samples." *Forensic Science International Genetics* 6, no. 3 (May 2012): 306–309. doi:10.1016/j.fsigen.2011.06.007.

Yang, Y., Xie, B., and J. Yan. "Application of next-generation sequencing technology in forensic science." *Genomics, Proteomics & Bioinformatics* 12, no. 5 (October 2014): 190–197. doi:10.1016/j.gpb.2014.09.001.

Zubakov, D., Boersma, A.W., Choi, Y., van Kuijk, P.F., Wiemer, E.A., and M. Kayser. "MicroRNA markers for forensic body fluid identification obtained from microarray screening and quantitative RT-PCR confirmation." *International Journal of Legal Medicine* 124, no. 3 (May 2010): 217–226. doi:10.1007/s00414-009-0402-3.

Zubakov, D., Kokmeijer, I., Ralf, A., Rajagopalan, N., Calandro, L., Wootton, S., Langit, R., Chang, C., Lagace, R., and M. Kayser. "Towards simultaneous individual and tissue identification: A proof-of-principle study on parallel sequencing of STRs, amelogenin, and mRNAs with the Ion Torrent PGM." *Forensic Science International Genetics* 17 (July 2015): 122–128. doi:10.1016/j.fsigen.2015.04.002.

Conclusions and Future Outlook of Next Generation Sequencing in Forensic Science

10

10.1 NGS Is Here

Since its introduction in the late 1990s, next generation sequencing (NGS) has found numerous applications in cancer and disease research and clinical applications, microbiology, and crop biology as well as bioforensics, biosurveillance, and infectious disease diagnosis (Kircher and Kelso, 2010, Yang et al. 2014, Budowle et al. 2014, Schmedes et al. 2016, Arenas et al. 2017, Minogue et al. 2019). Important lessons were learned from sequencing the human genome that led to sequencing an individual genome (Collins 2003). Upon commercialization of the sequencing instruments, clinical applications began and continue to increase. NGS has altered genomics research in the past fifteen years. Testing that was not affordable or technically feasible has been made possible by NGS (Patrick 2007, Mannhalter 2017). While many laboratories still use Sanger sequencing for forensic and clinical diagnostic applications, NGS is increasingly finding applications in these labs especially as the price for NGS has decreased to approximately $1000 (US dollars) per sample, making it more feasible for routine applications and casework (Mannhalter 2017). Full genomes are mapped with decreased cost and published almost weekly (Børsting and Morling 2015). NGS has demonstrated that it can overcome issues of efficiency, capacity, and allelic resolution presented by capillary electrophoresis and reduce the number of false positives in mixture analysis. While NGS is currently being used for the most challenging samples such as human remains samples recovered in cold cases, we expect it will be employed to analyze more routine samples in the future. Labs will decide how and where in their workflow NGS fits.

It has only been since the 2010s that NGS has begun to make an impact in forensic science (Minogue et al. 2019). NGS is currently being used for human identification, phenotyping, and ancestry applications using human blood, buccal, bone, or teeth samples (Jäger et al. 2017). The first NGS kits approved for collecting human genotyping data for Combined DNA Index System (CODIS) searches in the United States criminal justice system were only approved in 2019 (Verogen media release). In parallel to the progress in applying NGS to human genotyping applications, the tool has been found

DOI: 10.4324/9781003196464-10   145

useful in characterizing species of microorganisms for forensic applications (Minogue et al. 2019). While the research landscape in this area is still in its infancy, there are several trends and opportunities that are observed. As de Knijff wrote in 2019 in his paper, "From next generation sequencing to now generation sequencing in forensics," forensic use of CE also took time to be appreciated and adopted.

10.2 Why NGS?

As we have seen throughout this book, NGS has several advantages over CE approaches. NGS has been applied to diverse applications with successful results in each. NGS can be used in forensic casework to identify the source of a stain or biological evidence sample – even if it was from an identical twin. Sequencing STRs can illuminate differences that are masked in CE fragment analysis. NGS can be used to predict biogeographical ancestry and traits including eye color, hair color, and skin tone. SNPs that are multiplexed in NGS library prep kits with STRs are analyzed simultaneously instead of by multiple different assays. NGS can be used to determine the body fluid source of a biological sample, the site on the human body it was taken from, if the sample is human, and even which species are present in the sample without having to decide which species to test for *ab initio*. NGS has led to more complete DNA typing results from human remains including damaged, historic or ancient hair, bones, and teeth remains. NGS has led to more complete routine mitochondrial DNA typing and familial tracing. NGS has been used to determine the cause of death in an autopsy and signs of disease in oral and blood samples. It has been used to differentiate the source of hairs and cell phones and predict lifestyle patterns. NGS has enhanced the capabilities in bioforensics and biosurveillance. NGS has been used to differentiate soils and sites and found use for geolocation. It has been used in archeological studies to identify gut bacteria and infection. NGS can capture subtle differences between bacterial communities including foodborne pathogens and bioterrorism agents in samples without reliance on target genetic marker systems (Sjödin et al. 2012). NGS has also been applied to sequencing drug and "legal high" species including the opium poppy and marijuana to characterize simple sequence repeat (SSR) genetic markers and chloroplast genome and STR sequences, respectively (Celik et al. 2014, Oh et al. 2015, Houston et al. 2018).

The MiSeq and Ion series instruments are easier to use and maintain than CE. The commercial Converge and Universal Analysis Software systems are easy to use and intuitive. In our lab, NGS has been found to lead to more data, higher sensitivity, and increased precision than older methods with similar sample input (unpublished data). An advantage of NGS is the scalability and ability to look at data for many markers simultaneously in a sample. NGS can simultaneously type STRs and SNPs and detect isoallelic variation (Berglund et al. 2011). NGS can capture sequence differences for STR alleles of the same length (homoplasy) and polymorphisms in the flanking regions (Børsting and Morling 2015). NGS can better detect heteroplasmy in mitochondrial DNA typing. These features can aid in the individualization of samples and mixture deconvolution. NGS can be used to differentiate identical twins and perform age prediction (Weber-Lehmann et al. 2014, Silva et al. 2015, Daunay et al. 2019). The increased number of loci increases the discriminatory power. This is essential when the quantity of DNA is limited, and there is not enough DNA to perform a half dozen or more separate panels or when working with samples from closely related individuals. When compared to implementing separate aSTR, Y-STR, X-STR, and SNaPshot tests, NGS is time- and cost-effective. NGS is more cost-effective per base than Sanger sequencing for large numbers of samples and loci as the sequencing is performed in a massively parallel configuration.

With the identity and ancestry informative SNPs probed using the NGS kits, all samples can produce investigative leads even if the STR profile fails to make a hit in a database such as NDIS. Additional loci can be used in statistical calculations. NGS can also reveal errors that are not clearly resolved with CE including stutter caused by DNA strand slippage, base-pair errors not fixed by the polymerase during editing, slippage especially at homopolymeric strands, and errors as a results of sequencing miscalls.

10.3 Ongoing Challenges of Adopting NGS for Forensic Investigations

After years of development, NGS has demonstrated great promise for forensic casework. Although researchers and analysts have developed, piloted, and validated methods and kits, NGS remains a new tool that is being used in forensic casework. Even with all of the possibilities and advantages that NGS offers, there are still very real challenges that must be overcome for NGS to be widely adopted for forensic use.

Issues that need to be resolved prior to adoption for casework include nomenclature, data storage, funding, training, statistics, genetic privacy concerns, contamination issues, reporting, sample tracking, accreditation, casework needs, and acceptance by court (Alonso et al. 2017). The nomenclature for NGS-based STR alleles needs to be standardized. There is no convention for reporting isoalleles or accepted procedure for performing statistical evaluations of the newly identified alleles. Statistics need to be developed and uniformly employed to analyze new NGS alleles. Just as when any new technology is introduced, standard operating procedures (SOPs) need to be developed and the instrument and method need to be subjected to internal validations.

Beginning at the crime scene, investigators need to know the power and limitations of NGS in order to collect the appropriate samples or decide which samples are most promising to send to the lab. Other issues include the potentially long and complex chain of custody and speed of analysis if NGS is used (Gilchrist et al. 2015) and defining the analytical threshold (AT) to avoid overinterpreting the data (Young et al. 2017). As an example, to further analyze raw NGS data and noise to define the AT, FASTQ files were analyzed using the STRait Razor™ software and Python scripts instead of the Verogen UAS software (Young et al. 2017).

An issue that needs to be addressed with regard to NGS is cost. More competitors are needed in the industry to drive down the costs of forensic DNA analysis. Labs need to invest in and implement more automation to reduce preparation time and inter-operator variability. Centralized labs may help counties, states, territories, and countries achieve the economies of scale needed to make NGS cost-effective rather than cost-prohibitive. While Sanger sequencing is still the optimal method in terms of time and cost for sequencing short targets and pyrosequencing is ideal for probing DNA methylation variants, NGS technologies have several advantages when many loci need to be sequenced for each sample or for sequencing complete genomes or chromosomes.

Funding must be made available in the form of grants or included in allocated annual budgets to facilitate access to NGS, either in the form of new a local capability or for sample out-processing. The US DNA Capacity Enhancement and Backlog Reduction Program which funds grants totaling $82 million a year has been approved for the purchase of laboratory equipment, and reagents as well as training for systems that are approved for use with NDIS; grants can help reduce direct costs to labs seeking to implement new technology such as NGS (Verogen media release).

Even if funding is granted through a special program or allocated by states in the annual budget (Funding information for U.S. Forensic Laboratories), labs still must decide which NGS platform to adopt and validate the new kits and instruments at their labs. Labs need to develop and validate SOPs and workflows to process samples and store the high resolution and large sequence datasets (Gilchrist et al. 2015, Aly et al. 2015, Clarke et al. 2017, Mannhalter 2017). As some of the kits are sequencer-specific, the lab will need to decide upon the kit and sequencing instrument. The commercial NGS kits that have been introduced and are approved for input into CODIS are reliable and robust. The Promega PowerSeq 46GY system is sequenced on a MiSeq and the Verogen ForenSeq Signature Prep kit amplicons are sequenced using the MiSeq FGx. The Applied Biosystems Precision ID system amplicons can be sequenced using a ThermoFisher Ion series instrument. Qiagen's kits are compatible with the MiSeq. After sequencing is complete, the data analysis begins. Verogen sells its Universal Analysis Software (UAS) for analyzing STR and SNP data and a different version for analyzing mitochondrial DNA typing data. ThermoFisher's Torrent Suite and Converge software can be used to analyze data from its Ion series instruments. Qiagen's CLC Genomics Workbench can analyze and visualize data from all major NGS platforms. Versions are available for Windows, Mac OS X, and Linux platforms. Methods still in development will have to demonstrate that they are also sensitive, specific, reliable, and robust through developmental validation.

Many of the commercial software packages developed for NGS data analysis are hosted on the cloud which could be susceptible to outages and cyberattacks. Verogen's UAS and Illumina's BaseSpace applications are cloud-based software that can be accessed from any computer using a virtual private network (VPN) client and a lab can make unlimited accounts for its users. ThermoFisher's Torrent Suite Software can be accessed via the local area network.

Another challenge is the quantity of data produced. NGS generates a huge quantity of data. The data output from the MiSeq is approximately 1 GB per sequencing run. Labs must consider data storage options including external hard drives, cloud storage, and internal server storage when adopting NGS technology. Whereas forensic labs routinely maintain paper backup files of CE and quantitative data with NGS, it is no longer feasible to print hard copies of all of the data. While the actual sequence data files are not large in storage size, the raw digital photographs recorded after each base is added are cumulatively substantial in size. For example, the server shipped with the Verogen UAS can store approximately one hundred sequencing runs. A lab could choose to save only the original raw data and final analyses as intermediate data interpretation files can be reconstructed, as needed, using the software. Labs will need to establish which data to save and back-up and whether off-site services will be acceptable. Adoption of cloud storage is an option for storing a copy of all of the data that will be collected or in-house servers can be purchased and maintained to house the data. All of these options require additional infrastructure and funding, and supporting an in-house server may require additional IT support. Labs will also face the question of which sequencing output files need to be stored indefinitely.

NGS reporting and implementation guidelines have been released and continue to be rolled out. A new version of SWGDAM was released on January 12, 2017, that included NGS in the Internal Validation Guidelines for the first time (SWGDAM). On April 23, 2019, an NGS Addendum to the SWGDAM Autosomal Interpretation Guidelines included background information, core elements, interpretation guidelines, mixture interpretation guidelines, and a comparison of references and statistical weight (SWGDAM). The application of NGS is included in the FBI Quality Assurance Guidelines and Standards that became effective July 1, 2020 (FBI).

NGS is extremely sensitive. This is a great strength of NGS but also can lead to mixture profiles from samples contaminated with environmental DNA. Since NGS is much more sensitive than previous kits and methods, the quality assurance program needs to ensure that all products the lab uses from tips to tubes and other consumables are DNA-free otherwise a plant worker's non-relevant DNA could be typed. Scientists must be able to differentiate between mutation and error, especially in mixture samples. Error reads typically occur infrequently while true alleles will result in tens of reads. The minimum number of reads under various conditions (e.g., mixtures) needs to be established (e.g., 10 or 50 or 100 reads for low-level contributor) for each locus and sample. Labs need to allocate more analysis time for mixtures.

NGS poses challenges in implementation. Implementing a technology such as NGS requires training of existing staff and validation of the new technology, reagents, kits, and writing new or amended SOPs (Budowle et al. 2014). While the DNA extraction and quantitation instrumentation are largely transferable, staff will need to be trained in NGS technology. While most graduates of Forensic Science Education Programs Accreditation Commission (FEPAC)-accredited programs are well-versed in STR DNA typing methods using CE, as of early 2021, they likely have not had training in NGS. Skilled and experienced analysts will need advanced training courses. They should be reassured that the commercial library preparation kits and sequencing manuals for forensic applications are easy to follow and that their skills are easily transferrable to performing the new protocols. Verogen and ThermoFisher offer training to labs who purchase their instruments and adopt their kits. Colleges and universities are conducting NGS research and adding NGS courses. FEPAC-accredited institutions including Pennsylvania State University, Sam Houston State University, and Towson University are training new scientists in NGS applications for forensic science and developing new NGS-related forensic biology methods. In 2019, Towson University added undergraduate and Masters-level courses in forensic science (FRSC 422 and FRSC 622, respectively) focused on NGS for both autosomal and mitochondrial DNA analysis (Elkins and Zeller 2020). Other institutions offer online NGS courses.

On May 2, 2019, the US FBI approved profiles generated using Verogen's MiSeq FGx Forensic Genomics System for upload to the National DNA Index System (NDIS). With support from a contractor, Ohio participated in a pilot study of the ForenSeq kit. Washington, DC and California forensic labs have adopted ForenSeq in their labs. To introduce NGS to the Washington, DC

Department of Forensic Sciences, Dr. Jenifer Smith utilized a contractor for implementation support to mitigate the burden on her staff. The Baltimore City Police Department laboratory obtained an Illumina MiSeq instrument with grant support and is validating ForenSeq for casework.

While processing and preparing samples for NGS requires a similar amount of time as CE, the library preparation steps are more time-consuming and the sequencing step takes much longer. Whereas STR fragment analysis on CE takes approximately twenty minutes a sample, and several capillaries are routinely run in parallel, an NGS run with a MiSeq FGx instrument must run to completion to obtain the data for the samples so while ninety-six samples take a similar amount of time as CE, the time required is standard, so for one sample it is prohibitive. Furthermore, remote labs are challenged with continuous power for long NGS sequencing runs (Minogue et al. 2019). Microbial community profiling of human body fluids, human body site and geographic locations, and evidence items needs to be accepted by courts. Technical and biological validation of the various NGS applications will be required before it can be adopted as a standard tool for use in casework and acceptance into courts (Kuiper 2016).

Microbial NGS methods especially lack standardization of targets and analysis approaches including databases for statistical analysis, especially when unknown and rare taxa are encountered for interpretation using limited published study data. Further development of bioinformatics tools and processed and referred databases are needed (Minogue et al. 2019). The bioinformatics methods need to be able to map and identify sequence variants by critically evaluating raw sequence data (Gilchrist et al. 2015). The data output needs to be presented in an actionable format (Gilchrist et al. 2015). Other NGS-centric issues include depth of sequencing, higher error rates than Sanger sequencing, reproducibility, sensitivity, AT-sequence bias, and large number of targets and markers (Gilchrist et al. 2015, Aly et al. 2015). More studies of robustness of the method are needed (Gilchrist et al. 2015, Aly et al. 2015).

Familial DNA searching has begun in jurisdictions that allow it but this poses privacy concerns. Similarly, there are ethical considerations to consider. The ForenSeq NGS panel contains loci that codes for unique traits, as opposed to solely "junk" DNA. GEDmatch was a database founded to help users use their genetic profiles to locate family members based upon similarities in the genetic makeup. Individuals can upload their DNA profiles from one of several personal genealogy services. Initially, GEDmatch gave users the option of opting out of other uses including investigations. Verogen recently purchased GEDmatch for use in cold case and other criminal investigations. Now users and potential family member matches must opt into investigative use; otherwise, their samples are protected from these searches and their

Figure 10.1 Summary of challenges of adopting NGS for forensic investigations.

privacy is maintained. There were concerns when users previously had to opt out that GEDmatch was turning all users into suspects. Nevertheless, database security breaches are an ongoing concern.

Figure 10.1 summarizes many of the issues that need to be resolved including allele naming, data storage, statistics, and acceptance by the courts.

In spite of the concerns and challenges, countries around the world are working to bring NGS to the forensics workflow due to its advantages. NGS has been used in casework and missing persons investigations.

10.4 Early Successes of NGS in Forensic Cases

Genetic genealogy has been employed in investigations. DNA profiles of non-offenders from commercial companies including 23andMe, Ancestry DNA®, and My Heritage DNA tests are being used in searches to support law enforcement. NGS data and websites such as GEDmatch and Family Tree databases as well as traditional history research methods using the United States Federal Census, state birth indexes, Newspapers.com Obituary Index, US City Directories, US Obituary Collection, US Social Security records, and church membership and baptismal records have proved to be valuable in their approach. As a recent case example, GEDmatch was used by law enforcement to solve the decades-long cases of the Golden State Killer (Selk 2018).

The ThermoFisher HID-Ion AmpliSeq™ Ancestry Panel was used in a forensic case involving a carbonized corpse (Hollard et al. 2017). The autosomal STR profile did not lead to a profile so NGS was used to determine the eye color and biogeographical origin of the deceased. The team also conducted Y typing and mitotyping. NGS did lead to more information but lack of a sufficient database for interpretation was a drawback. Xiao (2019) recently described the design and implementation of a large-scale high-throughput automated DNA database construction using NGS.

The first case using NGS in a trial in Dutch Courts in January 2019 (de Knijff). The case included a sample from a complex sexual assault with a minor contributor at less than 10% that of the major contributor in the STR profile. The STR repeats were reported, but the traditional CE method does not permit the determination of underlying sequencing information. The DNA sample was analyzed using traditional PCR and CE methods and generated a hit in their convicted criminal database. The case resulted in an acquittal because many of the minor contributor's alleles were in the stutter position of the major contributor's alleles. Upon appeal by the prosecution and reanalysis of the samples using the MiSeq FGx, the minor contributor was distinguished from stutter, and likelihood ratio statistics were performed on the results. Upon hearing the new data and analysis, the judge ruled that the defendant was guilty of sexual assault (de Knijff 2020).

NGS has also been demonstrated for use in paternity cases. In a study, DNA isolated from sperm cells of monozygotic twins and blood from one of the twins' children was typed using ultra-deep NGS. The researchers used VarScan 2 to analyze the sequence data for somatic mutations. Individualizing SNPs were identified in samples originating from the child's father that were not found in the father's twin (Weber-Lehmann et al., 2014).

The France National Police implemented a decision tree for deciding whether to analyze samples with NGS or traditional CE. In summary, if a sample is limited or degraded or if a Y-STR profile is needed, NGS using ForenSeq library preparation is supported (Alvarez-Cubero et al. 2017). If the DNA profile is urgent (<72 hours), PCR followed by CE is recommended. Their workflows include mini-STR typing, Y-STR typing, autosomal STR typing, and phenotypic SNP typing using CE and mitotyping using Sanger sequencing. They suggest NGS use in complex kinship cases, to identify a very minor contributor, to obtain a genetic profile from highly degraded DNA, and to deconvolute a mixture using possible isomutations. The French National Police used NGS in 2018 to analyze a 2011 cold case. The first analysis was performed using Identifiler and the two DNA extracts showed a mixture of the victim's DNA and that of a very minor male contributor. Using NGS, the profile at D2S1338 was determined to contain two different seventeen repeat alleles (isoalleles) which led to assignment of the minor profile.

To date, only a few cases processed using NGS have been presented in court; each jurisdiction will have to assess allowing the introduction of data produced with the new methods in accordance with the law. Forensic laboratories may utilize NGS for serious crimes and cold cases in the future although caseloads may preclude using DNA typing for all cases.

10.5 Summary

In summary, John Butler wrote in 2005 that "the future is always challenging due to unforeseen innovation." While NGS continues to challenge scientists and labs, NGS is here and providing new opportunities for analysis to solve crimes. The developmental validation of NGS for forensic applications has been published in peer-reviewed journals. NGS is being applied successfully to criminal cases, mass disaster, and missing persons forensic casework.

Questions

1. List five advantages of NGS over CE-based DNA typing methods for forensic applications.
2. Compare and contrast the advantages and challenges of implementing NGS in place of traditional DNA typing methods.
3. Is it easier for a lab to move to NGS-based DNA typing when the sample is limited or plentiful? Explain.
4. Is the time for NGS "now" or not? Justify your response with references.
5. Is there a "gold standard" for NGS? Explain why or why not.
6. List and explain five issues that forensic labs face in implementing NGS.
7. What is the biggest challenge facing labs considering implementing NGS? Support your answer.
8. Discuss ethical concerns regarding the use of DNA data.
9. Are there risks associated with including human sequencing data in databases for law enforcement use? Explain.
10. Explain how NGS can be used to solve forensic cases that were intractable with traditional STR typing approaches.

References

Alonso, A., Muller, P., Roewer, L., Willuweit, S., Budowle, B., and W. Parson. "European survey on forensic applications of massively parallel sequencing." *Forensic Science International: Genetics* 29 (March 2017): e23–e25. doi:10.1016/j. fsigen.2017.04.017.

Alvarez-Cubero, M.J., Saiz, M., Martínez-García, B., Sayalero, S.M., Entrala, C., Lorente, J.A., and L.J. Martinez-Gonzalez. "Next generation sequencing: an application in forensic sciences?" *Annals of Human Biology* 44, no. 7 (November 2017): 581–592. doi:10.1080/03014460.2017.1375155.

Aly, S.M., and D.M. Sabri. "Next generation sequencing (NGS): A golden tool in forensic toolkit." *Archiwum Medycyny Sądowej i Kryminologii [Archives of Forensic Medicine and Criminology]* 65, no. 4 (2015): 260–271. doi:10.5114/amsik.2015. 61029.

Arenas, M., Pereira, F., Oliveira, M., Pinto, N., Lopes, A.M., Gomes, A., Carracedo, A., and A. Amorim. "Forensic genetics and genomics: Much more than just a human affair." *PLoS Genetics* 13 (September 21, 2017): e1006960. doi:10.1371/ journal.pgen.1006960.

Berglund, E.C., Kiialainen, A., and A.-C. Syvänen. "Next-generation sequencing technologies and applications for human genetic history and forensics." *Investigative Genetics* 2, no. 1 (November 24, 2011): 23. doi:10.1186/2041-2223-2-23.

Børsting, C., and N. Morling. "Next generation sequencing and its applications in forensic genetics." *Forensic Science International: Genetics* 18 (September 2015): 78–89. doi:10.1016/j.fsigen.2015.02.002.

Budowle, B., Connell, N.D., Bielecka-Oder, A., Colwell, R.R., Corbett, C.R., Fletcher, J., Forsman, M., Kadavy, D.R., Markotic, A., Morse, S.A., Murch, R.S., Sajantila, A., Schmedes, S.E., Ternus, K.L., Turner, S.D., and S. Minot. "Validation of high throughput sequencing and microbial forensics applications." *Investigative Genetics* 5 (June 30, 2014): 9. doi:10.1186/2041-2223-5-9.

Butler, J.M. "The future of forensic DNA analysis." *Philosophical Transactions of the Royal Society B* 370, no. 1674 (August 5, 2015): 20140252. doi:10.1098/rstb. 2014.0252.

Celik, I., Gultekin, V., Allmer, J., Doganlar, S., and A. Frary. "Development of genomic simple sequence repeat markers in opium poppy by next-generation sequencing." *Molecular Breeding* 34 (February 6, 2014): 323–334. doi:10.1007/ s11032-014-0036-0.

Clarke, T.H., Gomez, A., Singh, H., Nelson, K.E., and L.M. Brinkac. "Integrating the microbiome as a resource in the forensics toolkit." *Forensic Science International: Genetics* 30 (September 2017): 141–147. doi:10.1016/j.fsigen.2017.06.008.

Collins, F.S. "The human genome project: Lessons from large-scale biology." *Science* 300, no. 5617 (April 11, 2003): 286–290. doi:10.1126/science.1084564.

Daunay, A., Baudrin, L.G., Deleuze, J.F., and A. How-Kit. "Evaluation of six blood-based age prediction models using DNA methylation analysis by pyrosequencing." *Scientific Reports* 9, no. 1, (June 20, 2019): 8862. doi:10.1038/s41598- 019-45197-w.

de Knijff, P. "From next generation sequencing to now generation sequencing in forensics." *Forensic Science International: Genetics* 38 (January 1, 2019): P175– P180. doi:10.1016/j.fsigen.2018.10.017.

de Knijff, P. "Case study: How next generation sequencing resolved a difficult case, leading to the first criminal conviction of its kind." *Verogen* 2020: 1–4. Accessed November 27, 2020. https://cdn2.hubspot.net/hubfs/6058606/Verogen-First- NGS-Court-Case-Study_Final_VD2019024_8.5x11-web.pdf?__hstc= 238609695.bed74b81cf4041e42adad16833ab8584.1576870704888. 1576870704888.1576870704888.1&__hssc=238609695.1.1576870704888.

Elkins, K.M., and C.B. Zeller. "What is the CURE for limited DNA? A forensic science course focused on NGS." *Journal of Forensic Science Education* 2, no. 2 (2020). https://jfse-ojs-tamu.tdl.org/jfse/index.php/jfse/article/view/31.

FBI. "Quality assurance standards for forensic DNA testing laboratories." July 1, 2020. https://r.search.yahoo.com/_ylt=AwrJ7Fu6cxBge58AkaVXNyoA;_ylu=Y29sbwNiZjEEcG9zAzQEdnRpZANBMDYxNV8xBHNlYwNzcg--/RV=2/RE=1611719739/RO=10/RU=https%3a%2f%2fwww.fbi.gov%2ffile-repository%2fquality-assurance-standards-for-forensic-dna-testing-laboratories.pdf%2fview/RK=2/RS=ZY.1vT5BZwWmyh4tC9UI4j9Nwpw-.

"FBI approves Verogen's next-gen forensic DNA technology for National DNA Index System (NDIS)." May 2, 2019. https://verogen.com/ndis-approval-of-miseq-fgx/.

"Funding information for U.S. Forensic Laboratories." March 27, 2019. Funding Information for U.S. Forensic Laboratories.

Gilchrist, C.A., Turner, S.D., Riley, M.F., Petri, W.A., and E.L. Hewlett. "Whole-genome sequencing in outbreak analysis." *Clinical Microbiology Reviews* 28, no. 3 (July 2015): 541–563. doi:10.1128/CMR.00075-13.

Hollard, C., Keyser, C., Delabarde, T., Gonzalez, A., Vilela Lamego, C., Zvénigorosky, V., and B. Ludes. "Case report: on the use of the HID-Ion AmpliSeq™ Ancestry Panel in a real forensic case." *International Journal of Legal Medicine* 131, no. 2 (March 2017): 351–358. doi:10.1007/s00414-016-1425-1.

Houston, R., Mayes, C., King, J.L., Hughes-Stamm, S., and D. Gangitano. "Massively parallel sequencing of 12 autosomal STRs in *Cannabis sativa*." *Electrophoresis* 39, no. 22 (November 2018): 2906–2911. doi:10.1002/elps.201800152.

Jäger, A.C., Alvarez, M.L., Davis, C.P., Guzmán, E., Han, Y., Way, L., Walichiewicz, P., Silva, D., Pham, N., Caves, G., Bruand, J., Schlesinger, F., Pond, S.J.K., Varlaro, J., Stephens, K.M., and C.L. Holt. "Developmental validation of the MiSeq FGx forensic genomics system for targeted next generation sequencing in forensic DNA casework and database laboratories." *Forensic Science International: Genetics* 28 (May 2017): 52–70. doi:10.1016/j.fsigen.2017.01.011.

Kircher, M., and J. Kelso. "High-throughput DNA sequencing-concepts and limitations." *BioEssays* 2, no. 6 (May 18, 2010): 524–536. doi:10.1002/bies.200900181.

Kuiper, I. "Microbial forensics: next-generation sequencing as catalyst: The use of new sequencing technologies to analyze whole microbial communities could become a powerful tool for forensic and criminal investigations." *EMBO Reports* 17 (2016): 1085–1087. doi:10.15252/embr.201642794.

Mannhalter, C. "Neue entwicklungen in der molekular biologischen diagnostik [German]." *Hamostaseologie* 37, no. 2 (May 2017): 138–151.

Minogue, T.D., Koehler, J.W., Stefan, C.P., and T.A. Conrad. "Next-generation sequencing for biodefense: Biothreat detection, forensics, and the clinic." *Clinical Chemistry* 65, no. 3 (March 1, 2019): 383–392. doi:10.1373/clinchem.2016.266536.

Oh, H., Seo, B., Lee, S., Ahn, D.-H., Jo, E., Park, J.-K., and G.-S. Min. "Two complete chloroplast genome sequences of Cannabis sativa varieties." *Mitochondrial DNA Part A* 27, no. 4 (June 24, 2015): 2835–2837. doi:10.3109/19401736.2015.1053117.

Patrick, K.L. "454 life sciences: Illuminating the future of genome sequencing and personalized medicine." *Yale Journal of Biology and Medicine* 80, no. 4 (December 2007): 191–194.

Schmedes, S.E., Sajantila, A., and B. Budowle. "Expansion of microbial forensics." *Journal of Clinical Microbiology* 54 (2016): 1964–1974. doi:10.1128/JCM.00046-16.

Selk, A. "The ingenious and 'dystopian' DNA technique police used to hunt the 'Golden State Killer' suspect." *The Washington Post* 2018. https://www.washington

post.com/news/true-crime/wp/2018/04/27/golden-state-killer-dna-website-gedmatch-was-used-to-identify-joseph-deangelo-as-suspect-police-say/.

Silva, D.S.B.S., Antunes, J., Balamurugan, K., Duncan, G., Alho, C.S., and B. McCord. "Evaluation of DNA methylation markers and their potential to predict human aging." *Electrophoresis* 36, no. 15 (August 2015): 1775–1780. doi:10.1002/elps. 201500137.

Scientific Working Group on DNA Analysis Methods. "Interpretation guidelines for autosomal STR typing by forensic DNA testing laboratories." Approved January 12, 2017. Accessed January 23, 2021. https://1ecb9588-ea6f-4feb-971a-73265dbf079c.filesusr.com/ugd/4344b0_50e2749756a242528e6285a5bb478 f4c.pdf.

Scientific Working Group on DNA Analysis Methods. "Addendum to 'SWGDAM interpretation guidelines for autosomal STR typing by forensic DNA testing laboratories' to address next generation sequencing." Approved April 23, 2019. Accessed January 23, 2021. https://1ecb9588-ea6f-4feb-971a-73265dbf079c. filesusr.com/ugd/4344b0_91f2b89538844575a9f51867def7be85.pdf.

Sjödin, A., Broman, T., Melefors, O., Andersson, G., Rasmusson, B., Knutsson, R., and M. Forsman. "The need for high-quality whole-genome sequence databases in microbial forensics." *Biosecurity and Bioterrorism* 11 (2012): S78–S86. doi:10.1089/bsp.2013.0007.

Weber-Lehmann, J., Schilling, E., Gradl, G., Richter, D.C., Wiehler, J., and B. Rolf. "Finding the needle in the haystack: Differentiating 'identical' twins in paternity testing and forensics by ultra-deep next generation sequencing." *Forensic Science International Genetics* 9 (March 2014): 42–46. doi:10.1016/j. fsigen.2013.10.015.

Xiao, L. "Designing and implementing a large-scale high-throughput Total Laboratory Automation (TLA) system for DNA database construction." *Forensic Science International* 302 (September 2019): 109859. doi:10.1016/j. forsciint.2019.06.017.

Yang, Y., Xie, B., and J. Yan. "Application of next-generation sequencing technology in forensic science." *Genomics, Proteomics & Bioinformatics* 12, no. 5 (October 2014): 190–197. doi:10.1016/j.gpb.2014.09.001.

Young, B., King, J.L., Budowle, B., and L. Armogida. "A technique for setting analytical thresholds in massively parallel sequencing-based forensic DNA analysis." *PLoS One* 12, no. 5 (May 18, 2017): e0178005. doi:10.1371/journal.pone.0178005.

Index

A

adenosine monophosphate (AMP) 22
AFDIL-QIAGEN mtDNA Expert (AQME) 106
AFDIL whole mitochondrial genome
    method 101
agarose gel 39, 40
age estimation 147
ALlele FREquency Database (ALFRED) 80
alleles 65–68, 72–73
    for Y-STRs analyzed 80–82
analytical threshold (AT) 62
ancestry 8, 68, 76
Anderson sequence 96
Armed Forces DNA Identification
    Laboratory (AFDIL) 98
ArmedXpert™, 78
autopsy 125–126
Autosomal STR Genotype Report 68

B

Ballantyne, J. 68, 95, 139, 140, 141
bead-based normalization (BBN) 100
biogeographical ancestry (BGA) 68, 76
biothreat surveillance 127
body fluid analysis
    epigenetics 137–140
    future of 141
    microRNA 140–141
    mRNA 139, 140
    traditional methods 137
Børsting, C. 28, 32, 36, 139, 145, 147
Bowtie 77, 128
Budowle, B. 33, 79, 95, 97, 107, 108, 117, 119,
    124, 125, 127, 128, 145, 147, 148, 150
Butler, J. 3, 4, 18, 20, 32, 96, 97, 98, 102, 154

C

Cambridge Reference Sequence (CRS) 96
capillary electrophoresis (CE) 6, 20
    alleles for Y-STRs analyzed 80–82

chain termination sequencing 13–14
cluster density 57–58
cluster generation 50
Combined DNA Index System (CODIS) 5,
    108, 145
computed population statistics 67
Converge and Universal Analysis Software
    systems 146
cytosine methylation 138

D

databases 69, 108–109; *see also specific types*
dbSNP 80
2'-deoxyribonucleotide triphosphate
    (dNTP) 14, 16
2', 3'-dideoxyribonucleotide triphosphate
    (ddNTP) 14, 20, 21
DNA extraction 32–33
DNA IQ™, 33
DNA quantitation 34–35
DNA sequencing
    chemistries used in 13
        chain termination sequencing 13–14
        by ligation 16, 18
        pyrosequencing 14–17
    detection techniques 17, 18
    for human identification 1
    massively parallel sequencing 23–25
    platforms 19–22
DNA typing, for human identification 2–7
double-stranded complex 50

E

Elkins, K.M. 4, 18, 32, 34, 98, 118, 128,
    139, 150
epigenetics 137–139
Erasmus server, phenotype analysis using
    74–77
European Standard Set (ESS) 5
Exome-Seq analysis 78

F

Fastq files 58, 78
Federal Bureau of Investigation (FBI) 108
first-generation sequencing techniques
  pyrosequencing 21–22
  Sanger sequencing 19–20
  SNaPShot sequencing 20–21
fluorescence spectroscopy 34
fluorescent dye 17, 18, 51
fluorimetric quantification-based
          normalization 100
ForenSeq™, 8, 39–41, 60–63
ForenSeq mtDNA Control Region 98–99, 102
forensic biology 1, 2
France National Police 153

G

gene expression 137–138
GeneMapperID 21
genetic genealogy 152
Genome Analysis Toolkit (GATK) 78
genomics 25
geolocation 125–126

H

haplotype 102
Helicobacter pylori 129
Helicos BioSciences Heliscope 25
heteroplasmy 102
HID_SNP_Genotyper plugin 70–71
HID_STR_Genotyper plugin 71
HipSTR (Haplotype inference and phasing
          for Short Tandem Repeats) 79
HIrisPlex-S assay 74
HiSeq 23, 47
Holland, M.M. 95, 97, 98, 99, 101, 102, 107, 108
human genome sequence 1
Human Microbiome Project (HMP) 117,
          118, 120–121
  applications 121–124
human sequencing control (HSC) 41, 49,
          62–63

I

i5 and i7 index sequences 38
integrative genomics viewer (IGV) 71, 106
interpretation threshold (IT) 62
ion detection 19

  platforms 23, 24
Ion PGM Sequencing 53–54
Ion Reporter Server System 69
ion series instruments 146–147
ion series run failure, troubleshooting 92, 93
Ion Sphere™ Particle (ISP) Density 70
ion sphere particles (ISPs) 42

K

Kayser, M.H. 32, 68, 81, 95, 139, 141
Kidd, K. 35, 68, 95
de Knijff, P. 8, 146, 153

L

Lander, E. 138
Lednev, I. 137
ligation
  DNA sequencing by 16, 18
  platforms 24
luciferase 22

M

maintenance wash 52
Maq 77
massively parallel sequencing (MPS) 8, 23,
          141; *see also* NGS
  ion detection platforms 23, 24
  by ligation platforms 24
  for mitotyping in forensic testing 107
  Reversible Chain Termination MPS
          Platforms 23, 24
  single base extension platforms 25
  third-generation platforms 25–27
McCord, B. 108, 138, 139
messenger RNA (mRNA) 139, 140
5'-methylcytosine 138
microbial DNA profiling
  applications 129–130
  in archeology 129
  autopsy 125–126
  bioforensics and biosurveillance 127–128
  bioinformatic approaches and tools
          126–127
  geolocation 125–126
  infectious disease diagnostics 128–129
  lifestyle analysis 125–126
  NGS methodology in 119–120
  postmortem interval 125–126
  sampling and processing 118–119

microbiome analysis 117
microRNA (miRNA) 140–141
MiSeq FGx 58–59, 146–147
  instrument failure 87–89, 93
  run failure 89–93
MiSeq Test Software 88
mitochondrial chromosome 1, 96–98
mitochondrial DNA (mtDNA) typing 95
  for forensic applications 107–108
  methods 98
  sequence data
    and databases 108–109
    interpretation and reporting 102–107
  using next generation sequencing 98–101
MITOMAP database (MITOMAP) 97
Mixture Ace 78
modern six-dye kit 9
MPS *see* massively parallel sequencing
    (MPS)
My-Forensic-Loci-queries (MyFLq) 106–107

N

National DNA Index System (NDIS) 5,
    87, 108
next generation sequencing (NGS)
  alleles for Y-STRs analyzed 79–82
  body fluid analysis 137–141
  challenges 147–152
  data analysis 57–58, 79–80
  denaturation 41–42
  DNA extraction 32–33
  DNA quantitation 34–35
  early successes of 152–153
  for forensic DNA typing 8–10
  instruments 25, 28
  Ion PGM Sequencing 53–54
  library preparation 35–39
  library purification and normalization
      39–41
  microbial applications of 117–130
  mitochondrial DNA typing 95–109
  for mixture interpretation 78–79
  multiplexing 41–42
  sample handling and processing 31
  sample preparation process 31, 32
  sequence analysis software 77–78
  ThermoFisher Ion Torrent™ Sequencing
      53–54
  troubleshooting 87–93
  validation and applications 80–81
  Verogen MiSeq FGx' Sequencing 47–53

O

Organizational Scientific Area Committees
    (OSACs) 8
Oxford Nanopore instruments 25

P

Parson ISFG format 78
Parson, W. 79, 95, 98, 99, 101, 102, 107,
    108, 147
Perkin-Elmer Corporation 4
phasing 57
phenol-chloroform-isoamyl alcohol (PCI/
    PCIA) 32
phenotype 8
  characteristics 3
  estimation 68
  tertiary analysis 68
  using erasmus server 74–77
Phred score 58
polyacrylamide gel 39
polymerase chain reaction (PCR) 4
polymorphisms 2
postmortem interval (PMI) 125–126
post-run wash 52
PowerPlex™ 5
Precision ID mtDNA Control Region
    Panel kit 101, 106
PredictSNP 80
prephasing 57
Promega Corporation 4
Promega PowerSeq 78
Promega PowerSeq™ CRM Nested
    System 108
Promega PowerSeq™ 46GY kit 35, 36, 148
pyrosequencing 138–139
  detection techniques 19
  dispensation of nucleotides in 16
  DNA sequencing 14–17
  first-generation sequencing techniques
      21–22

Q

QIAcel graph, of PCR amplicon for 41
Qiagen's latest PyroMark Q48
    pyrosequencer 22
Q-Score 58
QualitySNPng 80
Quantifiler™ DUO kit 34
Quantiplex HYres Kit 34

R

random match probability (RMP) 67
real-time PCR methods 34
research use only (RUO) 47
restriction fragment length polymorphisms
        (RFLPs) 3
reversible chain termination MPS platforms
        23, 24
revised CRS (rCRS) 96
RFLPs *see* restriction fragment length
        polymorphisms (RFLPs)
RNAseq analysis 78

S

Sample Compare mode 106
Sample Genotype Report 68
SAMtools mpileup 78
Sanger sequencing 15, 19–20
Scientific Working Group on DNA Analysis
        Methods (SWGDAM) 8, 108,
        149–150
Scientific Working Groups (SWGs) 8
semiconductor sequencing 24
sequence alignment map (SAM) file 77
sequence diversity databases 79
sequencing by synthesis (SBS) 50
severe acute respiratory syndrome corona-
        virus 2 (SARS-CoV-2) 125
sex typing 4, 139
short tandem repeats (STRs) 4, 6, 69,
        79–80, 146
simple sequence repeats (SSRs) 4
single base extension platforms 25
Single-Molecule Real-Time (SMRT) 25
single nucleotide polymorphisms (SNPs) 6
single-stranded DNA (ssDNA) 22
small nuclear RNA (snRNA) 141
SNaPShot sequencing 20–21
SNiPlay 80
SNPdetector 80
SNPedia 80
SNPServer 80
snRNA *see* small nuclear RNA (snRNA)
spectroscopic methods 34
ssDNA *see* single-stranded DNA (ssDNA)
SSRs *see* simple sequence repeats (SSRs)
standard reference materials (SRM)
        98, 99
standby wash 52–53

STRbase 2.0 Beta 79
STRs *see* short tandem repeats (STRs)
STRScan 79
STRSeq 79

T

ThermoFisher 101
ThermoFisher Converge Software 69–74
ThermoFisher GlobalFiler™ 5
ThermoFisher HID-Ion AmpliSeq™
        Ancestry Panel 152–153
ThermoFisher Ion Chef™ robot 39
ThermoFisher Ion Reporter™ Software 69–74
ThermoFisher Ion Torrent™ Sequencing
        53–54
third-generation platforms 25–27
tissue-specific differentially methylated
        regions (tDMRs) 138
triplasmy 102
troubleshooting
        ion series run failure 92, 93
        MiSeq FGx instrument failure 87–89
        MiSeq FGx run failure 89–93
        NGS sequencing 87

U

ultraviolet light source 51

V

variable number of tandem repeats
        (VNTRs) 3, 4
Variant Analyzer BaseSpace app 105
Variant Processor app 103
Verogen ForenSeq™ Signature Prep kit 36
Verogen MiSeq FGx® 23, 24, 47–53
Verogen Universal Analysis Software 49,
        58–69, 103

W

WebLogo 80
whole genome amplification (WGA) 99, 119
whole genome shotgun (WGS)
        metagenomics 119

Z

Zeller, C.B. 150